Coral Reefs in the Microbial Seas

The Influence of

Fishing, Nutrients, Bacteria, Viruses,

&

Climate Change

on Nature's Most

Wondrous Constructs

Coral Reefs in the Microbial Seas

Forest Rohwer

with Merry Youle

Illustrations by Derek Vosten

PLAID PRESS

A DIVISION OF PLAID PRODUCTIONS, INC.

PLAID PRESS

Copyright 2010 by Forest Rohwer

All rights reserved. Published in the United States by Plaid Press, a division of Plaid Productions, Inc.

Plaid Press is a registered trademark of Plaid Productions, Inc.

Library of Congress has cataloging-in-Publication Data
Rohwer, Forest.
Coral reefs in the microbial seas: the influence of fishing, nutrients, bacteria, viruses and climate change on nature's most wondrous constructs

Includes bibliographical references.

ISBN: 978-0-9827012-0-1

Illustrations © 2010 Derek Vosten
Book design by Neilan Kuntz

www.PlaidPress.com

Printed in the United States of America
10 9 8 7 6 5 4 3 2 1

To my daughter Willow Segall
May you see the amazing things that I have

Forest Rohwer

To the coral holobiont
For reminding us that no one can make it alone

Merry Youle

For my sons, Ronan, Xavier, and Simon
Appearing in the dedication of a science book officially
makes you a dork for life. Enjoy it, never forget, all things
are fascinating, you just need to look at them close enough.

Derek Vosten

So, in the ocean…corals occupy places which more highly developed creatures could not fill. They form, as it were, the base of the great structure of animal life, on which the next higher forms rest; and though in the course of ages they may undergo some changes and diversification of form and structure, in accordance with changed conditions, their essential nature has probably remained the same from the very dawn of life on the earth.

Alfred Russell Wallace
Darwinism: An Exposition of the Theory of Natural Selection with Some of Its Applications. 1889.

CONTENTS

- less fish, less coral, more fleshy algae
- more microbes and more virus-like particles
- more potential pathogens
- more diseased corals

Christmas Atoll as a microcosm of all four atolls

Revisiting the Reefs

Anthropogenic stressors with local or global impact
Priority #1: Stop overfishing
- why are we fishing so much?
- the Tragedy of the Commons and ways around it
- fishing: legal and illegal

Priority #2: Stop nutrient enrichment
- short-term costs, long term benefits

Priority #3: Provide increased habitat protection
- marine protected areas and no take areas
- effective enforcement

Valuing coral reefs

PREFACE

Our intent in writing this book is to provide an accessible account of the splendor of coral reefs, to address why they are currently dying, and to offer some useful strategies for their conservation and restoration. We discuss how stressors resulting from human activities—overfishing, nutrient additions, and increasing atmospheric CO_2—are working together with the microbes to kill corals. The potentially devastating global effects of increased CO_2 capture the headlines, but our primary focus is on the local impacts (e.g., overfishing, nutrient enrichment) that are quietly killing corals today. Although coral reefs around the world are in trouble, this book is not all gloom and doom. It is our attempt to convey the beauty and excitement of coral reefs, not by spectacular photographs of that underwater world, but by sharing some of what we now know about the exquisitely intricate and interwoven communities that make up each coral reef.

Scientific research is an intensely social experience. You work closely day-in and day-out with the same people, often under stressful conditions. You share the adventure from start to finish—from soaring pie-in-the-sky ideas to the tediously precise and repetitive data collection and analysis, and lastly the endless stream of rewrites required for publication of your findings. Added to this, when doing field work you share sleeping quarters, dining area, and bathrooms for months at a time. This gives rise to a strong camaraderie, with all its associated bantering and generally good-natured arguing. To share this, we start each chapter with a glimpse into the research life, drawn from our scientific expeditions to the Northern Line Islands, the Caribbean, and other reefs of the world. The main characters are the scientists at work, classified here by the organisms they study: the Fish who, obviously, monitor the diverse fish populations; the Benthics who concentrate on the corals and the other organisms living in the benthos (i.e., on the bottom of the sea); and the lowly Microbes who observe all the life that is too small to be seen. (Be aware that these sections were written by a Microbe and betray his class prejudices.) These vignettes are intended to be engaging, a bit of dessert before each chapter. You can, if you prefer, skip them without sacrificing any of the serious book content.

Following these lighter beginnings, each chapter then adds a piece to the story of how coral reefs work and why they are having such a hard time today. The pieces fit together to build our main hypothesis that stressors resulting from human activities are favoring the microbes, to the detriment of the corals. This is an evolving story. The research is happening right now, and many details are yet to be worked out. However, we feel that enough is known at this point to speak out and provide specific recommendations for the preservation of coral reefs into the future.

Every effort has been made to keep the story short, something that can be read in an afternoon. Consequently, endless fascinating or significant details have been omitted from the main narrative; much other valuable research has been neglected. Some additional details can be found in the footnotes, and suggested sources for further reading are provided in the Appendices. Our hope is that the book, being focused on the big picture, will spark your interest in one area or another and then serve as a useful framework for your further investigations. There is much more to be explored.

Forest Rohwer & Merry Youle
May 2010

Coral Reefs
in the
Microbial Seas

INTRODUCTION

Sharks & Fish

The diver, one of the Fish, put the regulator into his mouth and slipped into the water. Moments later he seemingly levitated, SCUBA gear and all, back onto the gunwale of the dive boat.

"There are a LOT of sharks down there..."

This caused everyone to pause. The Fish, scientists that spent most of their lives underwater, were not known for their preservation instincts. In fact, they normally pestered sharks, sea snakes, and eels by swimming as fast as they could towards them. If one of the Fish thought there were possibly too many sharks, there must be a swarm of them below the boat. This posed a problem: how to get the data without getting eaten.

The only workable solution was to determine how hungry the sharks really were. With a slight nudge, one of the Microbes "helped" the Fish back into the water. His thrashing would either cause the sharks to fall back or they would be attracted to the frantic Fish and everyone lucky enough to still be safely on the boat would have a great view of shark feeding behavior. Hopeful eyes looked for blood clouds in the water. Did Spielberg get it right in Jaws?

After a couple of minutes it was clear that the sharks deemed the Fish simply too unappetizing to eat. This makes sense because Fish are generally allergic to soap and freshwater, a condition that makes an effective shark repellent. With a slightly disappointed shrug, the rest of the science team donned their SCUBA gear, flopped into the water, and entered a coral dreamland.

There were sharks everywhere. Fifty-plus grey reef sharks circled the boat and divers. Thousands of fish darted in and out of view. And dominating the scene was a massive wall of coral extending up from the blue depths. The Scientists were continually being distracted by the sea of life around them as they gathered their data. Giant vampire snappers bit SCUBA tanks and pencils. Then, as the dive was wrapping up and the Scientists were climbing back onboard, a distinct "squeak, squeak" was heard underwater.

"Melon-headed whales!" Everyone tumbled back into the water and swam off the reef into the deep blue. Soon they were surrounded by a pod of these dolphins-called-whales that seemed glad for some mammalian company in this vast expanse of ocean. As the people floundered awkwardly in the water, the short-finned dolphins zipped past seemingly showing off, and the sun dipped to the horizon.

Oooops. Sharks feed at night. The swim back to the boat was unnerving. The "lots of sharks" had turned into "my God there are lots of sharks." A larger shadow rose out of the depths—a ten foot tiger shark was coming up for his evening meal. A couple of pictures were snapped and everyone clambered back on board.

Scientists with years of breathing underwater were giggling. "Did you see that manta ray? He had to be 15 feet wing tip to wing tip." "That shark pressed his nose right up against my camera lens." The euphoria continued all the way back to the mother ship.

That night the Scientists ate a simple meal of pasta and red wine, avoiding the turkey sausages, as hundreds of sharks circled the ship. Every once in a while a flying fish would make the fatal mistake of entering the circle of light surrounding the ship, and the sharks would tear it apart. As the wine flowed, cutting-edge experiments involving the lowering of a Fish overboard were discussed.

Before long everyone drifted off to bed, wanting plenty of sleep before the morning. Two years of planning and work had finally gotten the expedition to a pristine reef and not a moment was to be wasted.

he scientists aboard the Line Islands Expedition—the Fish, the Benthics, and the Microbes—had journeyed to the middle of nowhere to solve a mystery. Beautiful coral reefs are dying around the world. Why?

Coral reefs as we know them have been around for 200 million years. The coral animals that build them belong to the phylum Cnidaria, a group that includes some of the most ancient multicellular organisms on the planet. Corals are not fragile creatures, unfit, destined for early extinction. Rather, they are extremely well-adapted and adaptable organisms that have survived dramatic environmental changes in the past. They aren't dying of natural causes. They are being killed. And we humans are the murderers.

What does a dead or dying coral reef look like? The deterioration is not always obvious, even to the experts. For starters, most of us don't realize what a pristine, pre-human coral reef *should* look like. Typically, the reefs we know and regard as healthy were already significantly changed by human activities decades—even centuries—ago. For example, tens of millions of large fish, sharks, sea turtles, and manatees were removed from the Caribbean starting in the 18th century. This slaughter severely wounded the coral reefs in that region. Nevertheless, it took two centuries for the damage to catch our attention in the form of coral death. This lag between cause and effect is frequently seen in ecosystems due to their *resilience*. A resilient ecosystem can recover from insults—up to a point—without visibly changing, thus blinding us to the damage inflicted.

There is another, more subtle reason for our blindness. Each generation of reef scientists, SCUBA divers, and fishermen tend to accept the state of a reef when they first behold it as normal. As the years go by, they compare their current observations against the way things used to be, the way—in their mind—they should be. When the next generation arrives on the scene, the reef has declined farther. What one generation regarded as degraded, the next considers normal. Furthermore, specific studies assessing the health of a coral reef typically compare it against a baseline drawn from the start of the current study or, at best, the observations of a previous study completed a few years earlier. Each new study starts with a more recent baseline, one that has already shifted a little farther from pristine. This phenomenon is called the *shifting baseline syndrome*. The historical shift in our conception of a healthy coral reef has been documented by Jeremy Jackson (Scripps Institution of Oceanography) and others.

Jeremy Jackson

One way to escape shifting baselines is to read the accounts left by the explorers who visited truly pristine reefs. When Captain James Cook first arrived at Christmas Atoll in the Central Pacific two hundred plus years ago, his navigator James Trevenen recorded, *"On every side of us swam sharks innumerable, and so voracious that they bit our oars and rudder...."* [a] Likewise, Captain Edmund Fanning, the American explorer who discovered two nearby atolls in 1798,

later recounted, *"...the sharks here are very numerous, and while the boat was on her passage into the bay, before she entered the pass, they became so exceedingly ravenous around her, and so voracious withal, as frequently to dart at, and seize upon her rudder and her oars...."* [b]

When visiting the same reefs in 2005, scientists were stunned to find literally no sharks. Not one was sighted in hundreds of hours of diving. Sharks are a hallmark of a healthy, productive coral reef ecosystem. For millennia, there were not only sharks on the reefs, but also groupers, snappers, trevallies, and other large predators, all in such abundance that we now find the accounts almost unimaginable.

The transition from vibrant, towering reefs, teeming with fish large and small, to a flat, algae-covered underwater landscape is often gradual.

---

[a] Collingridge, V. (2002). *Captain Cook: A Legacy Under Fire*, p. 313. The Lyons Press.

[b] Fanning, Edmund. (1833) *Voyages Around the World; with Selected Sketches of Voyages to the South Seas, North and South Pacific Oceans, China etc.* Collins & Haunay

Figure i-1: When arriving at Christmas and Fanning Atolls in the late 1700s, explorers were greeted by schools of sharks. Today, most of the sharks are gone, as well as many of the corals.

However, since the 1970s, the pace of decline has quickened as stressors increase and their effects compound. Even the Great Barrier Reef, one of the most protected in the world, is not exempt. This colossal structure has lost 20% of its coral during the last sixty years, and there is a noticeable decline in the large fishes, most prominently the sharks. In the Caribbean, SCUBA divers are disappointed to find that the once-flourishing, exotic undersea wonderland has vanished. Eighty percent of the Caribbean reef coral has died in the last thirty years; worldwide, 30% of the coral is severely damaged. The future? Scientists estimate that 60% of the world's coral could be irretrievably lost by the year 2030.

> "Every ecosystem I studied is unrecognizably different from when I started. I have a son who is 30, and I used to take him snorkeling on the reefs in Jamaica to show him all the beautiful corals there. I have a daughter who is 17. I can't show her anything but seaweed."
>
> Jeremy Jackson

To see beyond our shifted baselines, it is essential to study without delay the few remaining pristine reefs—including their sharks, corals, algae, and microbes.[c] One reef still in near-pristine condition is Kingman Reef, part of the Northern Line Islands. These "islands" are a cluster of atolls located just above the equator in the central Pacific. Kingman Reef and Palmyra Atoll, the northernmost two in the group, are both now well-protected possessions of the United States. Palmyra hosts a marine research station with a couple dozen reef-conscious personnel; Kingman is uninhabited because there simply is not enough land. To the south lie Fanning (Tabuaeran) and Christmas (Kiritimati), part of the Republic of Kiribati, and inhabited by about 2500 and 5000 people, respectively.

Kingman Reef is a pristine coral fairyland that provides researchers with one of the last remaining examples of a healthy reef. Here the

---

[c] Although *microbe* can be defined as any organism too small to be seen with the unaided eye, we use the term to specifically refer to the *Bacteria* and the *Archaea*. These groups are two of the three Domains of life, the *Eukarya* to which we belong being the third. All three lineages diverged very early in our evolutionary history. The two "microbial" lineages are distinct from one another although both are unicellular organisms characterized structurally by the absence of a prominent nucleus.

sharks rule, and the smaller fish hide in the shelter of the coral. The corals are thriving and colorful; large seaweeds are almost non-existent. Christmas Atoll lies only three hundred kilometers (a couple hundred miles) southwest of Kingman, but its five thousand inhabitants make it a different world for the coral. Adjacent to all of the villages the coral is dead or diseased. Sharks and other large predators are absent; only small fish remain to dart among the slime-covered coral rubble. This same devastation is occurring throughout the world wherever humans live next to reefs.

Kingman is a view into the past, to the time of Captains Cook and Fanning; Christmas is a glimpse into the future of all coral reefs if we continue on our current trajectory. In visiting these atolls, the research expedition to the Northern Line Islands could explore how coral reefs functioned in the past, as exemplified by Kingman Reef, and then sort out what is happening to them today, as illustrated by the devastation at Christmas Atoll.

The expedition carried on board the usual equipment for studying corals, fish, and other macroorganisms. In addition was an immense heap of apparatus and lab supplies for sampling and characterizing the microbes. Many of the tools used today to study microbes in the environment became available only within the past couple of decades. *As amazing as it might seem, until the 1970s, humans did not know that most of the life in the ocean is microbial. We now know that there are almost a billion microbes and 10 billion viruses per liter (approximately 1 quart) of seawater.* Coral reefs and the causes of their decline cannot be understood without considering these microbes. In fact, we will show that even though coral reefs are incredibly complex, microbes are the main determinants of coral reef health and decline. However, before we can do that, we need to get to the reefs, which is where our story begins.

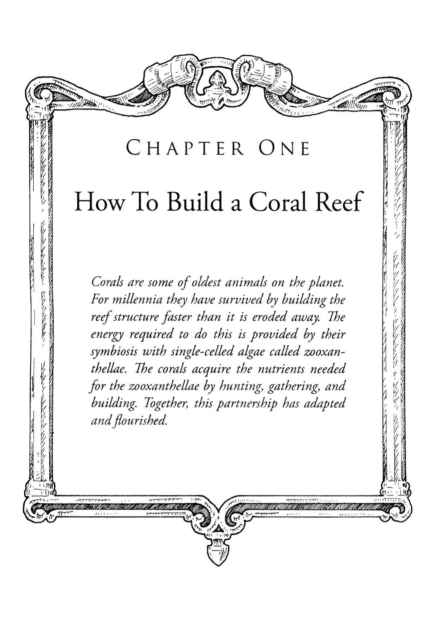

CHAPTER ONE

How To Build a Coral Reef

Corals are some of oldest animals on the planet. For millennia they have survived by building the reef structure faster than it is eroded away. The energy required to do this is provided by their symbiosis with single-celled algae called zooxanthellae. The corals acquire the nutrients needed for the zooxanthellae by hunting, gathering, and building. Together, this partnership has adapted and flourished.

I

Meeting the White Holly, Capt. Backen, & the Scientists

There is never enough money when planning a scientific cruise. This leads to choices. Should the expedition buy fuel, or food? Logic sends fuel to the top of the list. Fish can always be caught, but no one has yet come up with a reliable method for extracting diesel fuel from the open ocean. Fishing for food is a great strategy, provided someone brings fishing tackle or spear guns.

The first expeditional requirement is to secure a ship. Once the vessel has been contracted, the trip is sure to proceed; in fact, there is nothing that can stop it. Marine scientists will go to sea in the worst weather imaginable—if they have ship time. Everyone may be seasick, no work may get done because equipment cannot be deployed off a heaving deck. Yet, despite the misery, the contract is fulfilled and all participants are strangely satisfied.

The search for a vessel for the Northern Line Islands Expedition ended at the White Holly, the cheapest—and therefore the best—ship that could be found. In her youth, the White Holly served as a U.S. Navy yard freighter conveying ammunition to the warships in Pearl Harbor. Her middle years were spent in the service of the U.S. Coast Guard working the waters of southeastern Alaska and around New Orleans—until she was acquired by Captain Vincent G. Backen.

Captain Backen, like all ship captains, secretly wished that he did not have to share his ship with outsiders. His love for the White Holly was sullied by the necessity of hiring her out in order to support her expensive habits. The Holly's lines were not beautiful. Still Captain Backen loved her. In particular, he loved her engines, twin 353 Cats. They were immaculately maintained. And even the most impractical of scientists was aware that a ship's engines are sacred. In the middle of the Pacific Ocean, they must always work.

Accommodations on the Holly were spartan. The galley was too small for crew and scientists to eat together. Serviceable sleeping quarters and lab workspace were cobbled together. On the other hand, the Holly did have many useful features that made up for some of her charismatic shortcomings. Though large enough for the open ocean, her maximum draft was only ten feet. This meant that the Holly could, and did, maneuver into the shallow lagoons of coral atolls.

Most of the scientists on the expedition were senior, either PIs (Principal Investigators) or post-docs. Thus they had PhDs and were not to be trusted with truly important things, like the workings of the ship. Left to their own devices during the preparations, the scientists packed a few needed things and many superfluous others. First on their list was beer and wine. So greatly did they underestimate their needs that emergency libations had to be purchased for exorbitant prices from the one "store" within reach on Fanning Atoll. The scientists also brought along a shark cage that they mounted on the deck. There it looked most impressive and fueled secret dreams of chumming and tagging Great Whites. But the cage was soon filled with SCUBA tanks, then an air compressor was mounted against it. Thus anchored, the cage saw no active duty for the duration of the voyage.

Next on their list came the scientific equipment. Here a clear partitioning was apparent. There were four Fish on the scientific team and they were some of the best in the world at what they did—which was counting fish. They brought what was considered state-of-the-art technology for their chosen field: clipboards, pencils, and underwater paper. Often this sophisticated instrumentation did not work or components were misplaced. The Fish persevered through these setbacks and successfully counted fish on almost every attempt.

Just below them in the scientific hierarchy were the Benthics, those observers of the bottom-dwelling life forms, the inhabitants of the ocean benthos. They employed technology that was light-years ahead of the Fish. Amongst their gear were underwater cameras and PVC frames—vital supplements to the clipboards, pencils, and underwater paper. Their technological know-how far outstripped that of their fishy colleagues. They almost never mishandled their clipboard, pencil, or paper. They did, however, flood cameras with shocking regularity. In desperation, they rapidly commandeered unattended cameras on board the Holly. A watery doom awaited any camera that fell into Benthic hands.

Lowly, often forgotten, and almost left behind were the Microbes. Outnumbered and uncared for, their only responsibility was to measure and catalog everything that could not be readily seen, i.e., almost everything on the reef. Several tons of pumps, filters, microscopes, sampling devices, potentiometers, pipettors, and such were loaded into their on-board lab. An entire section of the ship's hold was filled with their plasticware and glass tubes. Their equipment was viewed as unnecessarily extravagant and its loading was met by much mutinous muttering amongst the counters-of-big-things.

Most disheartening to the Fish/Benthic mob was the RevCo, the mammoth –80 °C freezer that came along with the Microbes. It was securely lashed to one of the dividing walls in the ship's hold and provided with special power hook-ups. This worn hand-me-down was to be treated with the utmost respect at all times. Both the Crew and the Microbes watched it round the clock. Its task was to unfailingly maintain a temperature differential of greater than 100 °C between its inside and its outside, for two months, while being slammed around in the oceanic swells

After loading all of these essentials on the White Holly in San Francisco, the Scientists abandoned her. Her crew and captain would pilot her from there to Honolulu, Hawaii, and then on to Christmas Atoll in the Northern Line Islands. The Scientists, meanwhile, flew to Christmas via Duke's Restaurant & Barefoot Bar in Honolulu... not the best idea since seasickness compounded by a hangover is a really miserable way to start a voyage.

lthough he didn't start his voyage at Duke's Bar, Charles Darwin did spend much of his cruise puking over the side of the H.M.S. Beagle. Despite this slight problem, when he reached the South Pacific he was struck by the peculiar forms of the coral atolls. How, he wondered, did they develop their characteristic doughnut shapes?

Darwin, an observant geologist, came to the idea that each atoll had once been a high volcanic island encircled by corals. Year after year, the corals deposit more skeleton, extending the reef into deeper waters, eventually forming a ring-shaped barrier reef surrounding a volcanic cone. As more time passes, the island sinks (or in geological terms, subsides), and the coral keeps on building. By adding new skeleton aloft, they extend the reef upwards and maintain

Charles Darwin

their position within the sunlit *photic zone*.[a] Ultimately, the island disappears beneath the waves, leaving only a circle of reef teeming with life surrounding a blue lagoon—the archetypal atoll. The entire process may take a million years or more.[b]

Whereas atolls are formed around volcanic islands (Fig. I-1), similar processes occurring along continental coastlines or other islands give rise to barrier and fringing reefs. All of these massive coral structures form

---

[a] The *photic zone* is the upper layer of the ocean that receives enough sunlight to support photosynthesis.

[b] Darwin's novel notion—that volcanic islands and the ocean floor on which they sit are sinking—added fuel to his ongoing debate with the Agassiz family. The elder Agassiz, Louis, a geologist and zoologist, aggressively opposed Darwin's theory of evolution. Alexander Agassiz, his son, spent decades attempting to refute Darwin's notion of island subsidence.

The Darwin-Aggasiz atoll argument was settled in the 1950s when the U.S. Navy drilled several deep cores into Pacific atolls. They found that coral skeletons made up the bulk of the structure down to a depth of more than 4000 meters (13,000 feet). Since corals build reefs only in the photic zone, the only credible explanation was Darwin's basic theory. His insights into the geology of coral atolls was stunning. The theory of plate tectonics, which could provide a rationale for island creation and subsidence, was not fully developed until the 1960s. Darwin published his thesis in 1842.

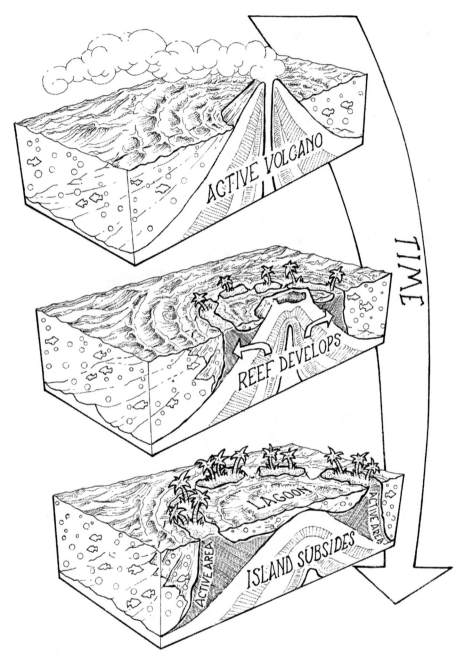

Figure I-1: The development of an archetypal coral atoll from a volcanic island. To survive, the corals must build new reef structure faster than it is eroded.

only when the water temperature is high enough for the corals[c] to build their skeletons quickly—fast enough to stay ahead of both erosion and the subsidence of the ocean floor. To maintain their place in the sun, coral reefs have to build (accrete) skeleton faster than it is eroded away. Net erosion spells death for the entire reef community.

Given enough time, coral reef structures can become epic in scale. The Great Barrier Reef covers hundreds of thousands of square kilometers and is easily visible from space. Nothing else built by living organisms, including humans, even comes close. Darwin never addressed how the humble corals build such massive structures. To explain this, we need to explore the daily activities of a coral colony.

ഗ § ഗ

Each colorful coral "head" is actually a colony made up of thousands of individual, but genetically identical, coral animals. An individual coral is called a polyp. Corals share their simple body plan with jellyfish, anemones, and other members of the ancient phylum Cnidaria. Evolutionarily, the Cnidarians and sponges are the oldest animals. However, Cnidarians exhibit numerous innovations not found in the sponges, including two organized tissue layers, more specialized cell types, and a nervous system. The two-layered body plan is unique to the Cnidarians; all the later animal lineages possess a third tissue layer.

Having been around for at least 540 million years, Cnidarians have survived environmental upheavals that are unfathomable to us, including all of the previous mass extinctions, which collectively wiped out 99.9% of Earth's earlier species.[d] Their tenacity speaks for itself. Corals are extremely tough and adaptable creatures.

What does a coral polyp look like? It is essentially a small version of the more familiar sea anemone, also a Cnidarian. Picture a soft sac, only

[c] The term "*coral*" encompasses a group of related but diverse organisms with different lifestyles and environmental requirements. Throughout this book, we focus on the reef-building corals. These are colonial stony corals (scleractinians) that live in shallow tropical waters and that house symbiotic algae (zooxanthellae) inside their cells.

[d] We are excluding the Great Oxygenation Event that predated the evolution of corals. Coral animals first appeared in the Cambrian period (540 mya). The modern reef-building scleractinian corals evolved during the Triassic (240 mya).

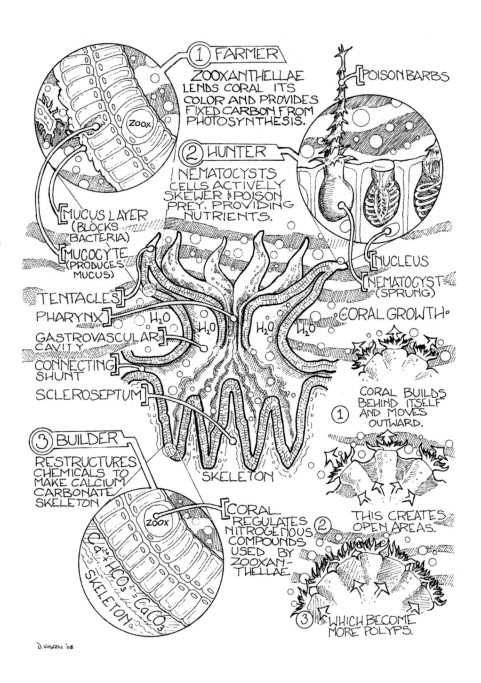

Figure I-2: Corals are remarkable animals that hunt for food with stinging cells, farm by raising algae, and build skeletons from minerals in seawater.

a few millimeters (tenths of an inch) in diameter, roughly cylindrical in shape, attached at its base to the rigid reef skeleton (Fig. I-2). The mouth is at the opposite end, encircled by tentacles that the coral animal can extend for capturing prey. The coral's tentacles and surface cilia sweep food into the sac where it is digested. A system of canals connects the sacs of neighboring polyps and allows the sharing of resources.

The walls of the sac are composed of two tissue layers. The inner layer, the endoderm, lines the internal cavity and functions in the digestion and absorption of food. The outer tissue layer, the ectoderm, interacts with the environment. It secretes the protective layer of mucus that covers the coral; its cilia sweep food particles toward the mouth and help clear sediment from the surface. The ectoderm also deposits the external skeleton as a cup-like chamber around each polyp. Each resident polyp can retract its whole body safely inside that skeletal shelter.

Together the ectoderm of many thousands of polyps produces the massive coral reef structures. These public works projects consume a lot of fuel in the form of sugars and other carbohydrates. To meet the demand, corals form a partnership with single-celled algae, called zooxanthellae,[e] which live as symbionts inside the cells of its endoderm.[f] The zooxanthellae are the coral's internal farm. Like most other algae, zooxanthellae carry out photosynthesis, using the energy in sunlight to convert CO_2 and water into energy-rich sugars and other carbohydrates. Essentially all of the carbon found in living cells in the form of proteins, carbohydrates, and other carbon-containing compounds came from CO_2 that was captured by photosynthesizing organisms.[g] Estimates are that the zooxanthellae transfer between 90% and 99% of their sugary production (photosynthate) to the coral. This infusion of energy is essential for the coral animal to grow and deposit skeleton fast enough to stay ahead of erosion.

---

[e] This partnership caused some consternation on the part of philosophers who sought to neatly divide the living world into two categories: plants and animals. Aristotle cataloged corals as plants. In the mid 1500s, Edward Wotton, the father of zoology, came closer to the truth when he gave them the enigmatic name Zoophyte (animal-plant).

[f] Bacteria and other unicellular organisms (such as the unicellular algae called zooxanthellae) that live as symbionts *inside* the cells of their host are sometimes referred to as *endosymbionts*.

[g] The biological conversion of atmospheric CO_2 into carbon-containing compounds such as sugars is called *carbon fixation*.

Being a successful farmer entails certain responsibilities. The coral must provide everything the zooxanthellae need for their photosynthesis. High on the list is the right amount of sunlight. When the seafloor sinks or the sea level rises, the corals must build the reef structure upwards fast enough to keep their zooxanthellae in the sun.

Since the zooxanthellae are cloistered inside coral cells, it's also up to the coral to provide them with the necessary nutrients. By "nutrients" we mean the compounds that contain nitrogen (as nitrites, nitrates, or ammonium), phosphorous, iron, or other elements that the zooxanthellae require for their intense rate of photosynthesis. Paradoxically, while typically located in nutrient-poor (oligotrophic) regions of the oceans, coral reefs have some of the highest levels of photosynthesis to be found in any marine environment. So how do the corals pull this off?

One way is by hunting the nutrient-rich animals that swim across the reef. Corals are armed with toxin-bearing harpoons on their tentacles. These sophisticated weapons, called nematocysts, are poised and ready to fire at all times.[h] When prey, such as a small, fast-moving crustacean known as a copepod, brushes against a nematocyte's protruding trigger, the cell fires and the barbed point delivers an immobilizing toxin. The tentacles sweep the now-powerless victim to the coral's mouth and into the gastrovascular cavity where it is digested.

Smaller prey, like microbes, are also packets of accumulated nutrients for the coral and their zooxanthellae. Because they are very small, microbes have very large surface-to-volume ratios. This makes them particularly efficient at taking up dissolved nutrients from the water. To capture microbes, corals secrete nets of mucus on their surface and then use their surface cilia to sweep the nets with the entrapped microbes into their open mouth. They also feed on microbes indirectly by hunting protists[i] that, in turn, graze on microbes.

Some of the nutrients the coral acquires by hunting and mucus netting are passed on to their zooxanthellae—like applying fertilizer to your

---

[h] These weapons are ancient. All of today's Cnidarians are descended from an early ancestor that possessed nematocysts. This is reflected in the name for the group, *Cnidaria*, which is derived from the Greek word for stinging nettle: *knide*.

[i] *Protist* is an informal term encompassing a very diverse collection of small organisms, many of them unicellular, but excluding the microbes. Protists include motile and drifting forms, photosynthesizers and hunters, and even some pathogens.

garden. The coral is then nourished by the abundant photosynthate produced by the zooxanthellae.

Corals also tap into the nutrient cache found in the deep ocean. Deep waters have high concentrations of nutrients because, for eons, marine organisms have been dying and their bodies sinking towards the seafloor. As their corpses descend, they decay, releasing nutrients into the water. When this happens in the photic zone, the nutrients are immediately grabbed by photosynthesizers and other organisms. But when the decay happens farther down, out of the sunlight, the nutrients tend to accumulate. In order for these nutrients to become accessible to the zooxanthellae or other photosynthesizers, they have to be "upwelled" back up to the photic zone. Upwelling regions occur naturally whenever oceanic currents collide with underwater structures such as islands and coral reefs. The front of the reef slows the incoming water, which effectively creates a vacuum on the opposite side. Deep water moves up to fill this vacuum, bringing the nutrients with it. Thus, by building reefs, corals create upwelling zones that bring nutrients to them and their zooxanthellae. The zooxanthellae can then produce more sugars and the coral can build yet more reef.

This symbiosis[j]—where the corals supply nutrients and shelter to the zooxanthellae and the zooxanthellae in turn contribute energy-rich sugars to the corals—is essential for building coral reefs fast enough to stay ahead of erosion. The entire process is finely-tuned for life in nutrient-poor tropical waters. Extra nutrients added by humans, i.e., from anthropogenic[k] sources, can disrupt this balance. Other factors such as warmer temperatures and more intense sunlight also jeopardize the coral-zooxanthellae symbiosis. Given enough environmental stress, the partnership falters. In the next chapter we'll explore one way that human activities are disrupting the symbiosis, a phenomenon known as coral bleaching.

---

[j] The term *symbiosis* has taken on numerous distinct meanings. We use it here to mean a close, prolonged association between two or more organisms of different species that may, or may not, benefit each member.

[k] *Anthropogenic* denotes caused or produced by humans.

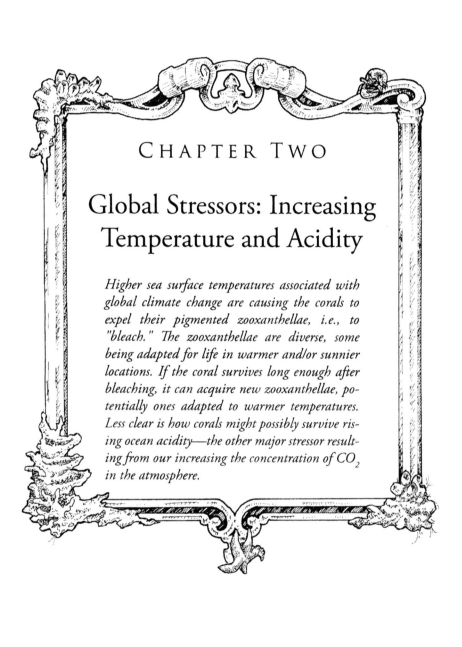

CHAPTER TWO

Global Stressors: Increasing Temperature and Acidity

Higher sea surface temperatures associated with global climate change are causing the corals to expel their pigmented zooxanthellae, i.e., to "bleach." The zooxanthellae are diverse, some being adapted for life in warmer and/or sunnier locations. If the coral survives long enough after bleaching, it can acquire new zooxanthellae, potentially ones adapted to warmer temperatures. Less clear is how corals might possibly survive rising ocean acidity—the other major stressor resulting from our increasing the concentration of CO_2 in the atmosphere.

II

Sergeant Uncareful and Fixing the Engine

The Line Islands are about as far from anywhere as you can be and still be on the planet Earth. So, if you're out on the reef in a dive boat and the motor quits and you start to drift out to sea, it is really, really important to have a radio. Due to budgetary constraints, there were only two radios per boat—the bare minimum because if you do start to drift away and one radio doesn't work, a backup is greatly appreciated. The batteries for the radios needed to be recharged every night. This was, of course, impossible to coordinate. Sergeant Careful, the dive officer on the first half of the Northern Line Islands Expedition, tried different strategies to make sure that this was done and that said radios were in the boats when the Scientists left the White Holly each morning. However, training PIs to perform simple tasks that will save their lives is nearly impossible. After assigning this task to various members of the scientific team and not getting the desired results, Sergeant Careful finally hit on a solution: do it himself and complain bitterly.

After keeping the Scientists alive for the first half of the trip, Sergeant Careful left the ship to undergo psychiatric treatment. He was replaced by another dive officer, initially known as Sergeant Uncareful, who was a hell of a lot more fun. No longer did anyone ask irritating questions such as, "Did you remember to refuel the boat?" No longer the never-ending sarcastic comments such as, "You could probably breathe better underwater if you connected the regulator to the air tank."

Yes, Sergeant Uncareful was much more fun. On his second day aboard the White Holly, he took one of the dive boats for a quick spin around the lagoon with the Most Distinguished PI aboard. Later both were spotted waving to the people on the White Holly, obviously having a great time exploring the reef. Everyone waved back, comfortable in the knowledge that the dive officer would call in using one of the trusty radios if they needed anything.

A couple of hours later, the boat with Sergeant Uncareful and the Most Distinguished PI had moved towards the horizon. Someone on board the White Holly commented, "They must be having the time of their lives. They didn't even take lunch and they're still out there."

More hours passed beneath the blazing tropical sun.

"Maybe someone should radio them and ask when they're coming back."

"That's funny, all the radios are here...and—oops!—here is the extra gas for their boat."

Upon retrieval, Sergeant Uncareful and his distinguished passenger were lethargic and didn't seem at all upset. "Why didn't you come get us when we waved?" one asked as he drifted off for a little nap.

A bit later, revived by sleep, drink, and food, there issued forth a torrent of questions. "Why wasn't there extra gas in the boat? Why weren't there any radios in the boat? Are you all idiots?" And on and on...

Sergeant Uncareful miraculously transformed into Sergeant Supercareful, and now he just wouldn't quit. More questions followed. "Where is the chase boat?" The idea was that when you called on the radio (assuming you had one) because you had run out of gas (who would be dumb enough to do that?), there would be a second boat that could come and rescue you.

"Tied to the top deck." The Scientists smugly imagined that Stupid Supercareful thought they hadn't remembered a chase boat.

"Why isn't it in the water?"

"Because the motor doesn't work."

Supercareful experienced a brief spell of Tourette syndrome: *"The fuckin' motor fuckin' better get fuckin' fixed fuckin' fast..."*

This was answered by very quiet mutterings. *"I liked Sergeant Careful better."* Obviously Supercareful wasn't cut out for life at sea with nineteen scientists on a budget. Only a couple of days and already he was starting to crack.

There are no repair shops in the middle of the ocean, which is one of the main reasons most marine biologists work only with devices constructed of PVC pipe and duct tape. The Fish, in particular, were completely petrified of complicated *"machines and electronics and stuff."* But they were good at lifting heavy objects and other sorts of manual labor, such as pulling the starter rope on a boat motor. They were also adept at pounding on a motor with a hammer, which they proceeded to do for a couple hours.

Having exhausted their technical know-how without solving the problem, the Fish coaxed one of the Microbes out of the lab. After some astute probing, the Microbe determined that the engine had *"frozen."* He then translated so that the Fish could understand: *"The stupid thing has rusted solid, probably because it hasn't been used for over a month."*

The easiest way to fix a frozen engine is to pour a little gasoline into one of the cylinders, then ignite the spark plug by pulling the starter rope. This is also the most dangerous method because the motor is just as likely to stay frozen and blow up. For survival, as well as efficacy, the amount of gas added to the cylinder needs to be precisely portioned. Even then, only a fool would actually be nearby when the starter cord was pulled.

The Microbe casually slopped some gas into one of the cylinders, screwed in the spark plug, and then moved behind the wheel house to avoid any shrapnel. "Hey Fishy, go ahead and pull."

Bam! The engine roared to life and the expedition was on its way once more.

The Microbe muttered, "Damn...not enough gas."

Of course, it is our casual use of too much gas and other fossil fuels that is sending more and more CO_2 into the atmosphere. This is not good for coral reefs.

he coral-zooxanthellae symbiosis is ancient. Fossil records document their liaison extending back at least as far as the Triassic—approximately 200 to 250 million years ago, a time when all of the continents were locked together in Pangaea. The algal partners, the zooxanthellae, are members of one of the most important groups of single-celled eukaryotes[a] in the oceans—the dinoflagellates. Their name comes from the Greek root *dinos*, to rotate or whirl, plus *flagellum*.[b] Most dinoflagellates are free-living and cruise around propelled and steered by two flagella. About half of them, including the zooxanthellae, carry out photosynthesis.

Dinoflagellates make the evening news when brief population explosions produce a visible algal bloom along an inhabited coastline. The bloom can color the water green, brown, red, or other shades, depending on which dinoflagellates are present. "Red tides" can temporarily shut down the local fisheries because the dinoflagellates contain toxins. When filter-feeders eat the algae, they accumulate the toxins in their bodies, and from there the toxins move on up the food on up the food web. Most toxins are not harmful to fish or shellfish, but eating an oyster that has eaten many toxic dinoflagellates can kill you. When a major algal bloom subsides, toxins are released into the water by the dying and disintegrating algae. Some of the toxin is aerosolized by wave action, and when inhaled by people can cause allergy or asthma-like symptoms.

One notorious toxin, ciguatoxin, is commonly found in fish in the tropics, especially in the top predators on coral reefs. The local population takes this seriously because ciguatera poisoning can cause long-term gastrointestinal and neurological effects, or even death. One of the main reasons that the vampire snappers (*Lutjanus bohar*) have not been completely fished out on many reefs is that people know which reefs have the toxin, and they know better than to eat snappers from there. (Fig. II-1)

Unlike their free-living, fun-loving dinoflagellate cousins, zooxanthellae live a cloistered life inside the cells of corals, giant clams, and

---

[a] *Eukaryotes* are organisms, unicellular or multicellular, that are characterized by complex intracellular structures and organelles, e.g., nuclei, chloroplasts, and mitochondria. All organisms belong to one of the three Domains of life: the Eukarya, the Bacteria, and the Archaea.

[b] The *flagellum* (plural: *flagella*) of dinoflagellates is a long, threadlike structure anchored inside the cell and extending through the cell membrane to the outside. Its propeller-like motion moves the cell through the water.

*Figure II-1: Vampire snappers (*Lutjanus bohar*), along with the sharks, are the top predators on coral reefs in the Line Islands. These fearless clowns of the reef are a bit on the "slow side." They habitually nip divers; anything left on the reef is immediately picked up and chewed.*

other marine invertebrates. All zooxanthellae belong to the genus[c] *Symbiodinium*. Although some types can be removed from their coral hosts and grown as free-living dinoflagellates in the lab, it is not known for certain whether any normally live and reproduce outside their hosts.

There is much give-and-take in this symbiosis between the coral host and its resident algae. The coral provides the zooxanthellae with the CO_2, water, and suitable amount of sunlight needed for photosynthesis, but

[c] A *genus* (plural *genera*) is a taxonomic grouping one level above species that typically includes a group of closely related species.

they restrict their growth by limiting their available organic nitrogen. As a result, the zooxanthellae grow slowly—about fifty times more slowly than their free-living relatives. At the same time, they work harder, photosynthesizing at least ten times faster. This combination of slow growth and intense photosynthesis leads to far more photosynthate than the zooxanthellae can use. Almost all of it goes to feed their coral host.

Photosynthate is akin to junk food, being mostly simple sugars, but it provides essential energy for the corals. Between 20% and 45% of the photosynthate donated to the corals is spent to make the copious mucus on their surface—a vital layer that protects against both physical damage and the entry of would-be pathogens. Without the contributions of the zooxanthellae, there would be no corals reefs as we know them.

⁊ § ⁊

Today, we are worried about the future of this ancient partnership. The reason? In recent decades, many corals in all regions of the world have, on occasion, expelled their zooxanthellae, most often in response to warmer water temperatures. This reaction is called bleaching because, divested of their pigmented zooxanthellae, coral polyps are transparent and their white skeletons show through. Without their in-house supply of photosynthate, corals lack the resources required for growth, reproduction, and reef construction.[d] Reduced mucus production renders the coral more susceptible to damage and disease. Although some bleached corals rebuild their zooxanthellae population and recover, many die in a matter of weeks, often due to opportunistic pathogens that are able to cause disease in the weakened corals. ***This combination of disease susceptibility and arrested reef construction makes bleaching one of the most serious threats to coral reefs today.***

Several circumstances can trigger bleaching: unusually high or low

---

[d] Depending on the species and the severity of bleaching, both the number of corals producing gametes (eggs and sperm) and the number of gametes produced will be reduced. Both of these factors impact reproduction. Since coral gametes are released into the water, fertilization success requires a huge number of gametes released by corals in close proximity to one another.

Baird, A.H., and Marshall, P.A. (2002) Mortality, growth and reproduction in scleractinian corals following bleaching on the Great Barrier Reef. *Marine Ecology Progress Series* **237**: 133-141.

Szmant, A.M., and Gassman, N.J. (1990) The effects of prolonged "bleaching" on the tissue biomass and reproduction of the reef coral *Montastraea annularis*. *Coral Reefs* **8**: 217-224.

water temperatures, unusually high sunlight intensity, reduced salinity, excessive sedimentation, disease, and various pollutants.[e] Prior to the 1980s, there had been only one large-scale coral bleaching event reported, and that one was in response to a sudden drop in salinity. It occurred in Jamaica when a hurricane dumped an exceptional amount of rain on the island and excess freshwater spilled onto the reefs. Many corals died afterwards. Since then, wide-scale bleaching has become a frequent occurrence in many regions of the world. That such events often correlate with unusually high water temperatures strongly suggests that coral bleaching will be even more frequent and more severe in the years ahead.

The link between the combustion of fossil fuels and coral bleaching is strong. The industrial revolution launched our prodigious burning of fossil fuels which has significantly increased the concentration of CO_2 in the atmosphere and contributed to global climate change via the greenhouse effect. Between 1861 and 2000, the average ocean temperature increased 0.4 – 0.8 °C. Temperatures will continue to rise along with further increases in atmospheric CO_2.[f] Even conservative estimates based on current trends and the mildest greenhouse scenarios predict an increase of 1–2 °C in sea surface temperatures by the end of this century. The actual shift will likely be much more dramatic.

A degree or two may not sound like a disaster, but most corals live close to their thermal limit. Thus, even a small temperature rise can still damage the corals. There is a rough rule-of-thumb: *If the maximum water temperature is 2 °C above the average summer peak temperature for that location for one week, or 1 °C above the average peak temperature for two weeks, the corals will bleach.*[g] Thus, it is not surprising

---

[e] Many questions remain unanswered regarding how these stressors disrupt the coral-zooxanthellae symbiosis. One hypothesis views this symbiosis as a controlled infection of coral cells by foreign algae. The "invading" zooxanthellae are thought to temper the coral's immune response so that they are accepted by the coral rather than destroyed. When the zooxanthellae are stressed by temperature or UV, they produce more reactive oxygen species or ROS. ROS are normal by-products of metabolism, but an excess of them disrupts cellular activities. Here, the increased production may prompt the host cell to now treat the zooxanthellae as foreign invaders and to activate its apoptosis or "cell suicide" pathway. Sacrificing a cell in order to eliminate an invader is one of the ancient strategies of the innate immune system in animals, including corals. When many coral cells respond this way, the coral bleaches.

[f] By 2050, atmospheric CO_2 is projected to be about double the pre-industrial level of 260 ppm. CO_2 concentration is expected to plateau near that level if we adopt the most stringent emission control plans now. If we continue with business as usual, atmospheric CO_2 will likely be three times the pre-industrial values by 2100.

[g] http://www.osdpd.noaa.gov/ml/ocean/coral_bleaching.html

that, on a global scale, increased coral bleaching correlates closely with increased atmospheric CO_2.[h]

The average sea surface temperature does not tell the whole story, as it is the local temperature at the coral polyp that matters. Corals do not live at the surface, but at various depths (most commonly less than 50 meters or 16 feet) in water mixed by currents to varying degrees. These factors may protect the corals in some niches. Similarly, average sea surface temperatures are indeed averages, with some regions being higher, some lower.

Of particular concern are climatic patterns, such as El Niño-Southern Oscillation events, that push local sea surface temperatures higher than the regional average. The most serious coral bleaching has accompanied such events. Approximately 99% of the corals in the Galapagos Islands bleached during the 1982–83 El Niño. Extensive coral death was brought to other reefs in the eastern Pacific, as well. The El Niño of 1997–98 was the strongest recorded to date, bringing 3–4 °C increases in some locales and unprecedented coral bleaching and mortality. For the first time in history, almost every coral reef in the world was affected. El Niño events are expected to increase in intensity and duration into the future—not good news for corals.

Recall that, in order for a coral reef to survive, the reef builders must add more skeleton than the reef loses to erosion. Obviously, when bleaching kills a significant fraction of the corals, reef construction slows. Even if the bleaching is temporary and the corals later recover their zooxanthellae population, construction is halted for weeks, even months, in the interim. One specific example for which we have both before and after data is the coral reef at Uva Island off the Pacific Coast of Panama. It lost

---

[h] Corals also live in precarious balance with another potential stressor: UV irradiation. Unfortunately, elevated sea surface temperatures and increased UV often hit at the same time as both are enhanced by reduced cloud cover and lighter winds. Ecosystems such as coral reefs located in the tropics receive more UV because, the sun being more directly overhead, the incoming sunlight passes through less atmosphere en route to the surface. To provide adequate light for their zooxanthellae, corals colonize sunlit sites where they are at greater risk for UV damage to their DNA and cellular membranes. Such damage can trigger bleaching.

Corals have multiple strategies for protecting themselves from UV. For one, they retract into the shade provided by their skeleton during bright daylight. For another, they synthesize a green fluorescent protein (GFP) that is thought to function as a finely-tuned sunscreen by absorbing damaging UV light and re-emitting it as less-damaging, lower energy visible light. Other sunscreens that are made by the zooxanthellae and transferred to their coral host include a family of unusual amino acids called *mycosporines*.

about half of its coral cover due to the 1982–83 El Niño bleaching event. Before that, the 2.5 hectare reef had been adding skeleton at the rate of 8,600 kg (19,000 lb) of calcium carbonate per year. Afterwards, the reef was losing almost 5,000 kg (11,000 lb) per year.

Net erosion affects the entire reef community by reducing the sheltering niches available. Simply put, fewer crevices mean fewer fish. The nearby land is affected, as well, since smaller, eroded reefs offer less coastal protection against the incessant pounding of the sea. This protection, important during normal times, can be decisive during extraordinary events such as storms and tsunamis.

ભ § ભ

When a coral reef bleaches, there will usually be some scattered coral colonies that appear completely normal. It is even common to see a bleached section within an otherwise pigmented colony. Sometimes these patterns are due to local variations in temperature or light; other times they reflect differences in thermal tolerance from one coral to another.

At first these differences were attributed to the coral animals themselves. After all, corals are obviously diverse, consisting of many groups quite distinctive in appearance and lifestyle. In contrast, the genus *Symbiodinium*, to which the zooxanthellae belong, was long considered to contain only one species and all zooxanthellae were assumed to share the same physiology, including the same thermal tolerance. This changed in the late 1990s when Rob Rowan (University of Guam Marine Laboratory) and Nancy Knowlton (Scripps Institution of Oceanography and the Smithsonian Institution) used DNA profiling to show that this assumption was wrong. The *Symbiodinium* genus actually contains multiple lineages, called *clades*. Further studies with more sensitive molecular techniques found yet more diversity, identifying multiple "types" within each clade. [i]

Once it was realized that there are genetically distinct clades of zooxanthellae, a host of new questions came to mind. Do the clades differ

---

[i] Reviewed in van Oppen, M.J.H. (2007) Hidden diversity in coral endosymbionts unveiled. *Molecular Ecology* **16**: 1125-1126.

in their temperature tolerance or in their ability to acclimate to different light intensities? Do different clades prefer to live with different coral species? Can a coral species partner with more than one type of zooxanthellae at the same time? Can they swap partners?

Initial answers came quickly. The various zooxanthellae clades do differ in their light and temperature tolerances, and the pairing of corals and zooxanthellae is not random. Early studies suggested that each coral species prefers to associate with one zooxanthellae clade, and that each clade has a preferred depth (i.e., irradiance level). More data brought more subtleties. A coral colony can host multiple zooxanthellae types—sometimes even within the same polyp. In the Caribbean, two of the dominant coral species, *Montastraea annularis* and *Montastraea faveolata*, form symbioses with members of three clades: clades A, B, and C. Taking a closer look, these particular associations make ecological sense. *M. annularis* prefers to live in brightly lit, shallow water; there it hosts clades A and/or B. On the other hand, *M. faveolata* favors deeper water with lower irradiance levels, and there it hosts clade C. When found at intermediate depths, either coral hosts two or even all three types, with clades A and/or B present in the sunlit portions of the colony and clade C present in shaded regions.

This suggests that the zooxanthellae clades can redistribute themselves within a coral colony based on their preferred temperature and/or irradiance level. This was confirmed experimentally by moving colonies of *M. annularis* and *M. faveolata* with their symbionts to different reef environments. When colonies hosting mostly clade C from deeper locations were shifted closer to the surface where they received more sunlight, they partially bleached. Six months later, the upper portion of the transplanted colony, where irradiance was highest, housed predominantly clades A and/or B—the clades adapted to higher light levels. In similar experiments, colonies housing multiple clades with the usual pattern of distribution were toppled. Within six months the zooxanthellae had redistributed themselves in accord with their new sunlit or shaded locations.

That a coral colony can house several types of zooxanthellae with different environmental tolerances and that these zooxanthellae can redistribute themselves to accommodate changing environmental conditions could be a boon for the corals in this time of rapid climate change. This possibility, combined with observations of coral bleaching and recovery, prompted Robert Buddemeier (Kansas Geological Survey) and Daphne Fautin (University of Kansas) in 1993 to suggest that bleaching might be the coral's way of replacing less heat-tolerant symbionts with more toler-

Robert Buddemeier   Daphne Fautin

ant ones.[j] This interpretation, known as the *adaptive bleaching hypothesis*, has been hotly debated ever since.[k]

When envisioning how the corals might accomplish this symbiont switching, two possible mechanisms come to mind. The corals might pick up new, better adapted zooxanthellae from their environment.[l] This requires adoptable zooxanthellae to be available locally and the adult coral to be capable of bringing them in-house. The second method makes use of the mixed population of symbionts already housed by most cor-

[j] Buddemeier, R.W., and Fautin, D.G. (1993) Coral bleaching as an adaptive mechanism: a testable hypothesis. *BioScience* **43**: 320-326.

[k] Hoegh-Guldberg, O., Jones, R.J., Ward, S., and Loh, W.K. (2002) Is coral bleaching really adaptive? *Nature* **415**: 601-602.
  Jones, R.J. (2008) Coral bleaching, bleaching-induced mortality, and the adaptive significance of the bleaching response. *Marine Biology* **154**: 65-80.

[l] Lewis, C.L., and Coffroth, M.A. (2004) The acquisition of exogenous algal symbionts by an octocoral after bleaching. *Science* **304**: 1490-1492.

als.[m] Imagine that bleaching selectively expels the less heat-tolerant zoo-xanthellae, leaving the more heat-tolerant symbionts in place. If those remaining symbionts multiply, the coral could once again host the normal number of zooxanthellae, but now most of them would be more heat-tolerant.

Has bleaching been observed to be adaptive? To answer this question, researchers have been collecting data on the distribution of zooxanthellae clades before and after major bleaching events. In general, the baseline distribution of the clades correlates with the normal maximum water temperature. Heat-tolerant clades, such as Clade D, are more common in warmer regions, such as the Persian Gulf. The heat-tolerant clade D was also found to be more common in some regions after the world-wide bleaching that accompanied the prolonged El Niño-Southern Os-cillation event in 1997–98.[n] Specifically, in the Indo-Pacific there was a population shift from the heat-sensitive clade C to clade D. Similarly, the proportion of clade D symbionts also increased following the 2006 bleaching event at the Great Barrier Reef.[o] These findings are good news for the corals. Perhaps corals will be able to adjust relatively quickly to increasing temperatures by partnering with zooxanthellae that are already evolutionarily adapted to thrive in warmer waters.

Also encouraging for zooxanthellae swapping is the seeming flexibil-ity of the coral-zooxanthellae partnership. Although most corals acquire their zooxanthellae when very young, there may be some symbiont com-ing and going at all stages. Corals are always releasing some zooxanthellae to their environs—one of the ways they control the number of zooxan-thellae living inside their cells. The zooxanthellae population fluctuates with the seasons in some Caribbean corals, there being fewer during the warmest months, more during the cooler seasons.[p] This, too, suggests

---

[m] Rowan, R., Knowlton, N., Baker, A.C., and Jara, J. (1997) Landscape ecology of algal symbionts creates variation in episodes of coral bleaching. *Nature* **388**: 265-269.

[n] Baker, A.C., Starger, C.J., McClanahan, T.R., and Glynn, P.W. (2004) Corals' adaptive response to climate change. *Nature* **430**: 741.

[o] Jones, A.M., Berkelmans, R., van Oppen, M.J.H., Mieog, J.C., and Sinclair, W. (2008) A com-munity change in the algal endosymbionts of a scleractinian coral following a natural bleaching event: field evidence of acclimatization. *Proceedings of the Royal Society B* **275**: 1359-1365.

[p] Fitt, W.K., McFarland, F.K., Warner, M.E., and Chilcoat, G.C. (2000) Seasonal patterns of tis-sue biomass and densities of symbiotic dinoflagellates in reef corals and relation to coral bleaching. *Limnology and Oceanography* **45**: 677-685.

that rebuilding one's zooxanthellae population, either from within or by acquisition, might be a normal exercise. Either way, this capability could help corals shift from less heat-tolerant to more heat-tolerant symbionts as temperatures rise.

ev § ev

Increasing atmospheric CO_2 caused by burning fossil fuel is also making the world's oceans more acidic—another hazard to coral reefs. An estimated 30% of the anthropogenic CO_2 is absorbed by the oceans where it becomes part of the pool of dissolved inorganic carbon (DIC). The increased dissolved CO_2 shifts the equilibrium between the various compounds that make up the DIC. As a result, the sea water becomes more acidic and the concentration of carbonate ions decreases. Carbonate ions are required by corals, crustose coralline algae,[q] and other marine organisms for building their skeletons and shells. Faced with increasing acidity, corals may have increasing difficulty in building new skeleton, and their existing skeleton may be weakened—perhaps even start to dissolve.[r] The increasing ocean acidification that lies ahead will affect even the most remote coral reef ecosystems.[s]

The average pH[t] of the ocean has already decreased about 0.1 pH unit from pre-industrial values, a shift that corresponds to a 30% increase

---

[q] *Crustose coralline algae* are so named because they form hard, calcium carbonate-containing crusts on the reefs. These crusts play an important role in cementing the reef structure—one of several reasons why these algae are beneficial to coral reefs. It is the *fleshy* algae that, when growing too abundantly, can take over a reef and kill the corals. For a more complete description of reef algae, see pages 101-104.

[r] Maoz Fine (The Interuniversity Institute for Marine Science) and Dan Tchernov (Interuniversity Institute for Marine Sciences; Hebrew University of Jerusalem) took a direct experimental approach to see if corals can survive in acidic seawater. They subjected corals in aquaria to acidic seawater at pH 7.3 to 7.6. (The present ocean pH is 8.3 to 8.6.) Within a month, the coral skeletons had dissolved, leaving naked polyps attached to the substrate. The polyps not only survived without their skeletons, but grew to three times their normal size. A year later, when the researchers brought the pH in the aquaria back to normal, the polyps built new skeletons and reformed their original colonies. Thus, in a sheltered, controlled, and predator-free environment, coral polyps can survive decalcification. In a real-life reef situation, the outcome would likely be different.
Fine, M., and Tchernov, D. (2007) Scleractinian coral species survive and recover from decalcification. *Science* **315**: 1811.

[s] Buddemeier, R.W., and Smith, S.V. (1999) Coral adaptation and acclimatization: a most ingenious paradox. *American Zoologist* **39**: 1-9.

[t] *pH* is a measure of the acidity or alkalinity (basicity) of a solution. The pH of most solutions falls between 1 and 14, with pH 1 being a very strong acid, pH 14 a very strong base, and pH 7 a neutral solution. A more acidic solution contains a higher concentration of hydrogen ions. A decrease of 1.0 pH unit corresponds to a ten-fold increase in the hydrogen ion concentration.

in the concentration of hydrogen ions and a decrease in carbonate ions. It is estimated that this has decreased the rate at which reef-building corals build their skeletons (their rate of calcification) by 20%. Ocean pH is projected to decrease another 0.3–0.4 pH units by the end of this century. This much change in pH is predicted to reduce coral calcification rates to 40–60% of normal. This is a momentous change. Even if CO_2 emissions are drastically reduced, absorption by the oceans will continue well into the future due to the slow mixing of ocean waters.[u]

Increasing ocean acidification translates into slower reef construction and weaker reef structures, both of which would shift the balance toward net erosion. Under acid conditions, the reef builders produce less dense reef structures that are eroded more quickly by waves and bioeroders[v] alike. Weaker reef structures also suffer more storm damage, which leads to decreased coral cover, which further reduces reef construction.[w]

On the hopeful side, corals already are accustomed to some variation in pH. The local pH at the surface of each polyp fluctuates daily—as much as 0.8 pH units. During the day, the photosynthesizing zooxanthellae consume CO_2, thus decreasing local acidity and raising the local pH. At night it is just the reverse: there is no uptake of CO_2 for photosynthesis but the polyp continues to respire, and the CO_2 it produces lowers the local pH.

There is also natural variation in the ocean pH from one region to another, and corals do live and build reefs in even the more acidic regions. One such acidic zone lies in the eastern tropical Pacific. Here coastal upwelling brings the deep, cold water with its abundant nutrients and elevated CO_2 concentrations to the surface. The pH of the Galapagos reef waters, for example, is about 7.9, well below the usual ocean pH of 8.3 to 8.6. Studying reefs in this area may tell us something about how

---

[u] Caldeira, K., and Wickett, M.E. (2003) Anthropogenic carbon and ocean pH. *Nature* 425: 365.

[v] *Bioeroders* are the organisms that convert coral reefs into fine white sand. These creatures (such as fish, molluscs, worms, fungi, and others) bore, drill, rasp, and scrape the skeletal structure, either from within or without.

[w] A healthy coral-dominated reef possesses an intricate topography, rich in niches and channels, and with a high degree of topological "roughness." The greater the roughness, the more fish and the more different kinds of fish on the reef. These irregular shapes also slow the incoming water, thus allowing the reef community to snag more of the incoming nutrients. Since the incoming waves and currents dissipate more of their energy on the reef, less remains for battering the nearby coastline.

coral reefs will fare in other regions when faced with increasing acidification. When Darwin visited the Galapagos Islands, he noticed that the coral reefs were poorly developed. They are small, patchily distributed, and rarely extend to a depth of more than ten meters (thirty feet). They have now been found to have the fastest rate of bioerosion of any reefs studied to date. Combined, these observations suggest that these reefs are relatively young and are barely staying ahead of erosion.

Naturally acidic conditions are also found at Flinders Reef, but here the outcome for the corals is different. Flinders is an extensive coral reef system in the west Coral Sea about 120 km (75 miles) outside the Great Barrier Reef. The pH of the local reef water for the past three hundred years was read from coral cores and found to have fluctuated between pH 7.9 and pH 8.2 in step with the 50-year climatic cycle in the region.[x] The pH is lowest when the climate is characterized by decreased surface winds, thus a time of reduced ocean currents and less flushing of the reef waters. Without effective flushing, the CO_2 produced as a by-product of reef calcification accumulates, driving the pH lower. When the surface winds increase, the pH returns to the normal range. Despite this cyclical acidification, Flinders Reef shows normal growth.

The effects of acidification can be more dramatic, as witnessed by a natural long-term "experiment" in the Mediterranean.[y] Just offshore from Italy are thriving communities of non-reef-building corals, beneficial low-growing algae, and abundant sea urchins. Scattered in the region are underwater volcanic vents that release enough CO_2 to lower the local pH by 0.2 to 0.4 pH units. In these islands of acidity, the community is dramatically altered. There are far fewer sea urchins, the algae are mostly large seaweeds, and there are no corals. Might similar changes be in store for other coral communities as ocean acidification continues?

Although corals build the primary reef framework, it takes more than corals to build a solid coral reef structure. Coral skeletons are bound together by the secondary frame builders, especially by the crustose coral-

---

[x] Pelejero, C., Calvo, E., McCulloch, M.T., Marshall, J.F., Gagan, M.K., Lough, J.M., and Opdyke, B.N. (2005) Preindustrial to modern interdecadal variability in coral reef pH. *Science* **309**: 2204-2207.

[y] Hall-Spencer, J.M., Rodolfo-Metalpa, R., Martin, S., Ransome, E., Fine, M., Turner, S.M. et al. (2008) Volcanic carbon dioxide vents show ecosystem effects of ocean acidification. *Nature* **454**: 96-99.

line algae (CCA) that form a tough calcareous crust overlaying the coral framework. This crust contains a high proportion of magnesium calcite, a form of calcium carbonate that is more soluble in mildly acidic conditions than is the form (aragonite) used to build the shells of molluscs and the skeletons of coral. In their aquaria experiments, Ove Hoegh-Guldberg (University of Queensland) and colleagues found that the CCA are themselves particularly sensitive to ocean acidification, more so than are the corals. CCA treated with elevated CO_2 showed markedly reduced productivity and calcification—the balance even tipping from accretion to dissolution. Increasing ocean acidity may impact these important reef builders first.

The final critical step in reef construction is the cementing of the reef. This is the geological process whereby the porous skeletal structures are filled with fine sediment and then everything is bound together with a limestone cement that precipitates out of the seawater. On many reefs, the portion exposed to incoming waves is especially heavily cemented to withstand the incessant battering. It is this cementing that is hampered by the low pH in the eastern tropical Pacific, including the reefs in the Galapagos. Both cementation and skeletal dissolution are governed by the physical properties of the various forms of calcium carbonate (aragonite, magnesium calcite, and others). Therefore, reef structures are certain to weaken as the oceans become more acidic.

℘ § ℘

Increasing sea surface temperatures and acidification are the main global threats to corals. We call these stressors "global" because they will affect corals in all regions, albeit to varying degrees. Their full impact lies in the future, when atmospheric CO_2 has risen further. Right now, however, corals are being killed primarily by the local stressors that we discuss in Chapters V through VIII.

Looking to the future, it is impossible to predict exactly what the impact of increased atmospheric CO_2 will be on coral reefs in any particular location. There are many variables that can locally buffer or intensify the effects. We also don't know how corals and the other reef organisms will respond to the changes. The concentration of CO_2 affects many physiological processes, enhancing some while reducing others, and these reactions vary from one marine organism to another. That some corals are able to live in regions with naturally warmer and/or more acidic seawater suggests that others may be able to adapt rapidly to moderate changes.

We mentioned earlier that corals may be able to adapt quickly to rising water temperature by exchanging their resident zooxanthellae for ones better suited for the new conditions. A similar strategy may help the corals to cope with other environmental stressors. Such possibilities arise because a "coral" is more than a polyp plus zooxanthellae. There are other partners that might contribute flexibility, the most likely ones being the microbes that we will meet in the next chapter.

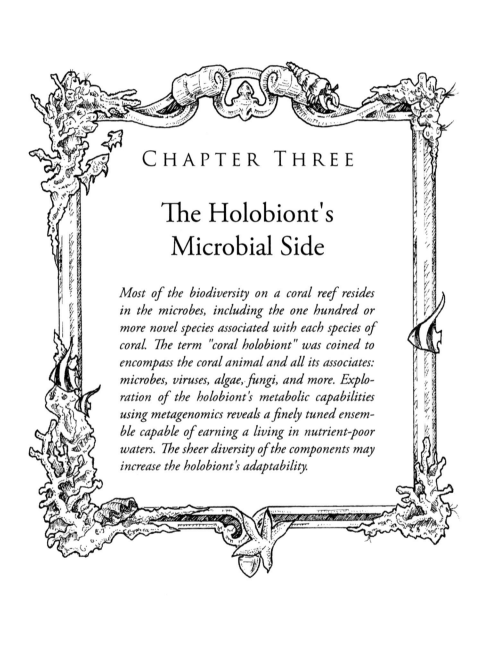

CHAPTER THREE

The Holobiont's Microbial Side

Most of the biodiversity on a coral reef resides in the microbes, including the one hundred or more novel species associated with each species of coral. The term "coral holobiont" was coined to encompass the coral animal and all its associates: microbes, viruses, algae, fungi, and more. Exploration of the holobiont's metabolic capabilities using metagenomics reveals a finely tuned ensemble capable of earning a living in nutrient-poor waters. The sheer diversity of the components may increase the holobiont's adaptability.

III

Clever Sample Labeling

When a scientific expedition sets out, possibly the single most important item on the agenda is the naming of sampling sites for future referencing. Using the boundless power of groupthink, the Scientists aboard the White Holly succeeded in developing a cumbersome and ultimately dysfunctional naming system. To begin with, it was decided that the first location, Christmas Atoll, should be referred to as Kiritimati—its modern name under the governance of the Republic of Kiribati. Thus, all the sites from that atoll would be designated KIR, instead of the expected CHR, or even the ubiquitous XMAS. Soon hundreds of samples with KIR labels on them were stashed in the faithful RevCo freezer at −80 °C—a temperature so cold that the future sorting of samples would burn the fingers. The labels were written by different people, most of whom had PhDs and consequently could no longer write legibly. Samples and yet more samples were crammed into the freezer until merely opening the door precipitated a cascade of tubes and baggies tumbling merrily to the floor. The KIR label became a major problem when the last atoll on the itinerary was designated KIN, for Kingman. Imagine differentiating between KIR and KIN on hundreds of frost-encrusted hand-scribbled labels as your fingers burn with cold and precious samples roll about on the floor.

Having successfully employed the best method for confusing samples from the first and last atolls, the researchers next came up with alternate designations for the same site. The torturous logic that led to this improvement was, thankfully, washed from mem-

ory by the imbibing of large quantities of alcohol. The legacy lives on, however, and to this day the same site might be designated as KIRB1 for Christmas Benthic site 1, KIRF3 for Christmas Fish site 3, or even KIRM2 for Christmas Microbe site 2. Subsequent scientific discussions went something like this:

Microbe: *There were about a million microbes per ml at M2. {Please note the social importance of dropping the island designation to demonstrate that one is in the know.}*

Fish: *I think there were lots of sharks there, too. Wasn't that F3?*

Benthic: *No, M2 is F2 and B6.*

Fish: *We didn't see any sharks at B6, but there were lots at F2.*

Microbe: *But aren't B6 and F2 the same?*

Fish: *Did you mean KIRB6 or KINB6?*

With the sample-naming system in place, the research teams were ready to get down to serious work. Two small boats were used to venture forth from the anchored White Holly. Gasoline for these crafts was stored on the stern of the Holly. To get it from there to the boats, the fuel was hand-pumped into portable tanks that were then carried up one ladder, shimmied past the navigation house, carried down another ladder, and thence lowered into the waiting boat. So each day began with someone hauling fifty pounds of fuel up and down ladders. Our good fortune: the oldest member of the scientific team was an inveterate masochist who readily took this job upon himself.

While gas was being transferred, the Microbes organized their gear: super-suckers, diver-deployable Niskin bottles for collecting water samples at various depths, and any cameras that hadn't been stolen by the Benthics. At the same time, the Benthics organized their ropes and rods for defining transects, PVC photo quadrats, and unlocked underwater camera housings, along with several hundred pounds of dive gear. Then everyone was ready to head out. About a half-mile later, everyone turned back to retrieve the clipboards, pencils, and underwater paper needed by the Fish, who tended towards confused. Finally, finally, the whole party was able to set out for the study site.

Each day, the Microbes dove once in the morning, returning to the White Holly by noon. At each site, eight water samples were collected at different depths. A small volume of water was filtered through an Anodisc—a glass wafer with holes so small that neither microbes or viruses could pass through. The microbes and viruses caught on the Anodisc were then stained so they fluoresced brightly and could be counted under a microscope. The fortunes of the coral—good health or decline—were written in these little pinpoints of light.

I n the late '90s and early '00s, I had the good fortune to work with two exceptional scientists at the Scripps Institution of Oceanography in San Diego, California. One was Farooq Azam, known for his pioneering work on the Marine Microbial Food Web.

Luckily, Azam was someone always open to new ideas. So when I approached him and proposed that we work together to study coral-associated microbes, he didn't point out that: (a) collectively we knew essentially nothing about corals, or (b) I might just be looking for an opportunity to go SCUBA diving in the tropics. Instead, he immediately proposed that we head downstairs to meet with Nancy Knowlton—the second of those exceptional researchers.

Farooq Azam   Forest Rohwer

Knowlton is a leader in coral reef science with a lengthy list of notable accomplishments. She discovered at least three cryptic species within the primary reef-building corals in the Caribbean, the *Montastraea* genus. Through this work she has been instrumental in making humanity aware that we are rapidly losing species through extinction that we didn't even know existed. It was Knowlton, working with Rob Rowan, who demonstrated that corals are host to numerous different types of zooxanthellae that are spatially distributed within a coral colony based on the local environment. This pioneering work added new dimensions to the field of coral-zooxanthellae symbiosis and the adaptive bleaching hypothesis discussed on page 47. When Azam and I approached her about studying the microbes on corals, Knowlton asked only one question: *When can you leave for Panama to start collecting samples?*

Nancy Knowlton

At the time when this work began, most previous studies of coral-associated microbes had been focused on pathogens involved in coral diseases and had relied on culturing[a] the microbes of interest. In the mid-1980s, John Paul at the University of South Florida had already shown that there were over 100 million microbes per square centimeter (0.16 square inch) of surface on healthy corals.[b] This is more than ten times as many Bacteria as are found on a typical patch of human skin or on the soil in a temperate forest. This large number also means that there are as many microbes on that square of coral surface as there are in the entire water column above it. This observation strongly suggests that microbes are important in some way to the coral. So, how do healthy corals interact with the billions of microbes that bathe their surface each day?

The first task for our trio was to determine what types of microbes live on healthy corals. This was not a simple look-see. Initial studies showed that there are far more microbes present than can be cultured using standard procedures. So, to study all the microbes, not just the small culturable minority, we needed to use methods that did not involve

---

[a] To *culture* a microbe is to multiply the organism under controlled laboratory conditions, typically in a nutrient broth or on a nutrient-containing agar plate.

[b] Paul, J.H., DeFlaun, M.F., and Jeffrey, W.H. (1986) Elevated levels of microbial activity in the coral surface microlayer. *Marine Ecology Progress Series* **33**: 29-40.

culturing. Fortunately, such culture-independent methods had been developed starting in the late 1980s. Paramount is the use of 16S rDNA[c] sequences to identify all the microbes—culturable and unculturable—from an environment. This methodology that had revolutionized environmental microbiology now provided us with a way to identify all the microbes from the mucus layer on the surface of corals. (In case you are unfamiliar with how 16S rDNA can be used to characterize microbes, we have provided an introductory tour in Appendix A.) *The initial survey based on 16S rDNA found that there are more than one hundred unique bacterial species associated with every species of coral. Half of the 16S rDNA sequences found on corals represent new bacterial species; one quarter are so different from all known Bacteria that the organisms could be considered members of new genera.[d]*

Furthermore, the coral-associated Bacteria aren't just the local neighborhood microbes casually stuck in the coral surface mucus. They are distinctly different from those in the water column above the coral and hence constitute distinct, coral-associated communities. This symbiosis between corals and their Bacteria can be quite specific. Corals of the same species growing at different locations have the same bacterial associates, even when the corals are located over a thousand kilometers (600 miles) apart.[e]

From the bacterial point of view, a coral reef is a complex and varied landscape, offering diverse niches for potential colonization. Each coral species provides a distinctive habitat, one that is hospitable to only some bacterial species. Even within a single coral colony, different regions house dissimilar Bacteria, as in the case of the branching coral *Porites furcata*. A particular bacterial species that colonizes the tips of the branches is not found in the mid-sections.

---

[c] *16S rDNA* is the gene that encodes a component of the ribosomes of microbes, specifically the ribosomal RNA found in the small subunit of the ribosome. Refer to Appendix A for a detailed explanation of how this gene is used to identify microbes.

[d] The concepts of *species* and *genus* came directly from careful observation of the multicellular organisms that make up the visible natural world around us. When we turn instead to the microbial world, these concepts are not easily applied, leaving us to resort to arbitrary—but nevertheless useful—definitions. Microbes whose 16S rDNA sequences are at least 97% identical are typically classified together as a species; those whose sequences are at least 93% identical are assigned to the same genus.

[e] Rohwer, F., Seguritan, V., Azam, F., and Knowlton, N. (2002) Diversity and distribution of coral-associated bacteria. *Marine Ecology Progress Series* **243**: 1-10.

Knowing that there are approximately a thousand coral species, each with over a hundred bacterial species that are found exclusively in association with them, we realize that there may be as many as 100,000 different bacterial species that live with one coral or another. When you consider that all of the other multicellular animals on the reef may have their own microbiotas,[f] the number of different Bacteria on a coral reef becomes staggering. Divers are struck by the diversity of the creatures they see swimming or floating past, as well as those crawling across or decorating the bottom. But most of the biological diversity on a coral reef is out of sight, found among the invisible microbes, the vast majority of which have not yet been studied.

∽ § ∽

So far we have seen that coral polyps live intimately with their zooxanthellae and their specific microbes.[g] Corals also host a number of other organisms, including fungi, protists, and viruses. Because all of these organisms are important components of a functioning coral colony, some years ago my colleagues and I proposed that this collaborative consortium be given its own name: the *coral holobiont* (Fig. III-1).

When one looks at a coral holistically as a coral holobiont, other insights and questions naturally follow. Is there flexibility in each of these partnerships? We have already seen that a coral species can partner with several different types of zooxanthellae and that the particular mix present is related to the local temperature and light level. Reasoning by analogy, one would suppose that a particular bacterial niche in the holobiont might be filled by any of a number of different Bacteria, depending on environmental conditions. If so, this would suggest that the holobiont might adapt quickly to environmental change by exchanging some of these other partners. To explore this possibility, we need to know not only who they are, but also what role these numerous and diverse organ-

---

[f] The *microbiota* encompasses all of the microbes normally harbored by a healthy organism, such as the flora living on our skin and in our gut.

[g] The microbes associated with corals also include a distinctive population of Archaea that differs from that in the surrounding water column. These Archaea are as diverse and as numerous as the coral's symbiotic Bacteria, but their relationship with their coral host is less specific. The same archaeal species may be found on many different species of coral.

isms play in the life of the holobiont. The zooxanthellae were easy—they provide energy—but very little is known about the ecological roles of the microbes, even less about the viruses and protists.

Most coral-associated microbes are known only by the presence of their unique 16S rDNA sequences. Fortunately, on occasion, even this limited information can provide clues as to their activities. Sometimes their 16S rDNA sequence is so similar to that of some well-characterized microbes that we know they are all closely-related members of the same group. If some members of that group are known to carry out photosynthesis, fix nitrogen,[h] or cause disease in plants or animals, we would predict that the unidentified members might play a similar ecological role. Often this is useful and accurate, but sometimes this approach leads to the wrong conclusion. All too often, closely-related microbes possess some distinctly different metabolic capabilities, and thus they carry out different functions in the community.[i] To get around this complication, we used another culture-independent method to assay the metabolic capabilities of the entire microbial community associated with a coral. This approach is called *metagenomics*.

Some definitions are needed before proceeding. The entire linear DNA sequence of a particular organism makes up its *genome*. A *metagenome* is simply an agglomeration of all the genomes present in a sample collected from a particular environment. As with 16S rDNA analysis, it is not necessary to culture the organisms. The basic procedure is to collect the microbial cells from the soil, a hot spring, the mucus on a

---

[h] *Nitrogen fixation* is the biological conversion of gaseous nitrogen (N_2) into ammonia (NH_3), which exists in solution as ammonium ions (NH_4^+). Only a few groups of Bacteria are capable of carrying out this reaction. Prior to the recent advent of fertilizers containing industrially fixed nitrogen, all other organisms depended on these nitrogen fixers, directly or indirectly, to meet their essential need for usable nitrogen.

[i] The rDNA genes are essential core genes required by all cells, including microbes. The 16S rDNA genes that are often used to identify microbes are inherited from one generation to the next and thus are almost identical in all members of a microbial species. In addition to such core genes, microbes typically carry other genes required for specialized activities or for taking advantage of some unusual opportunity. Such specialization genes can be acquired by a different mechanism. They often move from one organism to another of the same species—or even of a different species—via various mechanisms of horizontal gene transfer. As a result, members of the same species with identical 16S rDNAs might fulfill different metabolic roles in the community because they carry different specialization genes. A prime example of this is *Vibrio cholerae*, the cause of cholera. One can readily isolate *V. cholerae* from the environment, but most of them won't be pathogens. The pathogens among them cannot be distinguished from the non-pathogens by their 16S rDNA. The pathogens able to cause the disease carry the specialization gene that encodes the cholera toxin, a gene that can be acquired by horizontal gene transfer. In this way, studies that rely solely on the 16S rDNA to characterize microbes can miss some important differences.

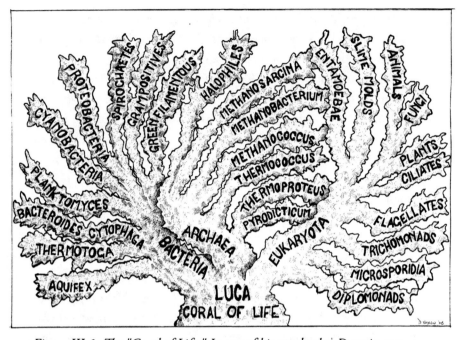

Figure III-1: The "Coral of Life." In one of his notebooks,[j] Darwin pondered that the "Tree of Life" representing the evolutionary relationships of all organisms could be better represented as a "Coral of Life." A coral colony would provide a more fitting analogy, one uniting his theory on the divergent evolution of species with Charles Lyell's theories on extinction. As Darwin knew, in a coral colony the living polyps are found only at the tips of the branches, the majority of the structure being dead. Analogously, the organisms alive today are perched at the tips of the branches of the evolutionary tree, while those at its base are extinct. In the end, the branching tree won out, likely because it was the better choice for conveying his message to a public enthralled at the time with genealogical family trees. Today, we think the "Coral of Life" particularly fitting because we now know that the coral holobiont comprises organisms from all three Domains of life. (LUCA is the Last Universal Common Ancestor from which all cellular organisms are descended.)

coral, or some other environment, extract their DNA, sequence as much of the DNA as possible, and compare the DNA sequences with known genes. This technique has led to numerous discoveries as researchers set

---

[j] http://www.articledashboard.com/Article/Darwin-s-Metaphor--The-Branching-Tree--Coral-and-the-Linguistic-Image/217246

to work exploring the microbial communities in one environment after another.

The hardest part of metagenomics is the data analysis that falls under the rubric of *bioinformatics*. This step is formidable because today's sequencing technology generates relatively short DNA sequences, each one referred to as a *read*. Sequencing of a typical metagenome sample yields from 50,000 to 500,000 reads. The bioinformatics starts by using computers to compare every read in the metagenome to the sequences of all known genes in the DNA databases. This step is, as the saying goes, computationally-intensive and can take weeks or months. The result will be a list of genes whose sequence is convincingly similar to that of a read in the metagenome. Some of those identified genes may encode a protein of known function, such as an enzyme involved in cellular metabolism.[k] In that case, it is likely that some organism present in the sampled environment is carrying out a similar function. For example, the sequences of numerous genes required to produce the light-capturing apparatus used for photosynthesis are known. Finding those sequences in the reads in a metagenome would indicate that the sampled community includes photosynthesizing organisms. Similarly, the presence of genes for virulence factors[l] indicates the existence of pathogens in the community. In this manner, *functional metagenomics* can profile the potential metabolic capabilities of the community as a whole—without the need to culture any of the organisms. This approach also makes it easier to detect those metabolic processes that are carried out by two or more community members working in tandem—interactions that can be overlooked when individual genomes are analyzed one by one.

Other useful insights can be derived by tallying the number of copies of each sequence found in the reads from a metagenome. These sequences come from the genomes of different organisms that were present in the sampled environment in different abundances. If there were one hundred of microbe X and ten of microbe Y for every single microbe Z, then

---

[k] An organism's *metabolism* is the sum of all the chemical reactions carried out by its living cell(s). These include the synthesis of molecules such as proteins and lipids, as well as the digestion or breakdown of other molecules, such as sugars, to acquire needed energy.

[l] A *virulence factor* is a molecule synthesized (and often secreted) by a pathogen that enables the pathogen to successfully infect a potential host.

the relative number of copies of their genomes in the metagenome will also be roughly one hundred, ten, and one. As a result, when we do the bioinformatics analysis, we would be one hundred times more likely to get information about microbe X than microbe Z. *Therefore, the relative abundances of different genes in a metagenome can tell us which organisms and which genes are most successful in that environment.* If the genes encoding the photosynthetic apparatus are highly abundant relative to others, this would indicate that the community is dominated by photosynthesizers.

The first metagenomic analysis of a coral holobiont was carried out by Linda Wegley Kelly (San Diego State University) when she investigated the microbial community associated with the mound coral *Porites astreoides.* To collect the microbes, Kelly crushed samples of the coral and used a centrifugation method to separate the microbes[m] from the larger coral cells. She then extracted and sequenced the microbial DNA, thus obtaining a microbial metagenome, also called a *microbiome.*[n] When the bioinformatics was finished, something totally unexpected emerged: a five-partner coalition that regulates nitrogen cycling in the holobiont.

Why is nitrogen cycling so important? Nitrogen is an essential nutrient for all organisms, being needed to build proteins, DNA, RNA, and other vital cellular constituents. Lack of usable nitrogen limits the growth of many organisms in the nutrient-poor waters where coral reefs are found. The exuberant and enduring success of reef-building corals is closely tied to their ability to meet their needs for nitrogen. For that purpose, they don't rely on just one strategy. As builders, corals encourage the upwelling of deeper, nutrient-rich water, thus bringing nitrogen compounds to them; they also acquire organic nitrogen[o] as hunters by feeding on plankton, and as gatherers by mucus netting microbes.

Another source of nitrogen for the coral holobiont is the abundant atmospheric gaseous nitrogen (N_2). No plant or animal, including coral

---

[m] These "microbes" included, in addition to the Bacteria and Archaea, some endolithic fungi (probably as spores), some viruses, and some mitochondria from the coral cells.

[n] A *microbiome* is a metagenome made up of microbial genomes.

[o] *Organic nitrogen* is nitrogen incorporated in cellular constituents such as proteins, DNA, RNA, and various metabolites. This is in contrast to inorganic nitrogen compounds such as nitrogen gas and various nitrates.

polyps, can convert gaseous nitrogen to organic nitrogen; however, the coral *holobiont* can fix nitrogen for its own use, courtesy of its sizeable population of nitrogen-fixing Bacteria. These in-house nitrogen fixers had been known for some time. The surprise in Kelly's microbiome was that nitrogen fixation by the holobiont is coupled with comprehensive nitrogen recycling carried out by several other partners.

Animals, including corals, produce ammonium (NH_4^+) as a metabolic waste product. Because too much ammonium is toxic, most marine animals get rid of it by excreting it into the water. The excreted ammonium is transformed through a series of steps, each one carried out by a particular group of microbes. The first two steps are the oxidation of ammonium to nitrite (NO_2^-) and then its further oxidation to nitrate (NO_3^-). This is termed *nitrification*. Another group of microbes converts the nitrate to nitrogen gas (N_2), a process called *denitrification* because the nitrogen is lost from the food web, escaping back to the atmosphere.

Kelly's analysis of the coral-associated microbiome showed that, instead of discarding the toxic ammonium, the coral holobiont treats it as a precious nitrogen resource and recycles it. Some of the ammonium is used directly by the zooxanthellae and the endolithic fungi[p] as their nitrogen source. Excess ammonium is converted to nitrate by a specific group of Archaea.[q] The nitrate is then converted back to ammonium by the endolithic fungi. The only part of the cycle missing from the holobiont is denitrification. Putting this all together, we see that the holobiont can use virtually all forms of nitrogen found in its environment (i.e., gaseous nitrogen, ammonium, nitrite, and nitrate) for the synthesis of the organic nitrogen compounds needed for its own growth and metabolism. Once acquired, the nitrogen is frugally recycled within the holobiont. Anthro-

---

[p] *Endolithic* means growing inside stony substances such as rocks and corals. Thus, *endolithic fungi* are fungi that live within the coral skeleton. Normally these fungi do not harm the corals, but when skeleton deposition is slowed, due to coral bleaching for example, the fungi can bore into the chambers housing the live coral polyps and feed on them.

[q] For some time it had been known that some Bacteria carry out nitrification. Relatively recently a group of marine Archaea were identified that also perform this transformation in the oceans. Soon thereafter, these Archaea were found to be associated with corals.

Beman, J.M., Roberts, K.J., Wegley, L., Rohwer, F., and Francis, C.A. (2007) Distribution and diversity of archaeal ammonia monooxygenase genes associated with corals. *Applied and Environmental Microbiology* 73: 5642-5647.

pogenic inputs of excess nitrogen can disturb this efficient operation.[r]

The nitrogen cycling carried out through the joint effort of these five partners in the holobiont—coral polyps, Bacteria, Archaea, zooxanthellae, and endolithic fungi—suggests an exciting possibility. Perhaps the particular organisms that make up each group vary depending on the local environmental conditions. If so, the slightly different holobionts so produced would, as a group, be able to thrive under more varied conditions than any single holobiont could.

Three of the groups–the Bacteria, the Archaea, and the fungi–possess far more varied metabolic capabilities than do any animals, including corals and humans. They have more diverse ways to harvest energy and are at home even in environments that appear to us to be inhospitable to life. Earlier we considered that the swapping of zooxanthellae might help corals deal with higher sea surface temperatures. Similarly, exchanging one microbial or fungal constituent for another with slightly different capabilities might enable the holobiont to cope with a wide range of conditions and stresses.

On the other hand, those diverse organisms that we classify as microbes also include pathogens with the nasty ability to kill corals. Thus, partner swapping in response to environmental impacts can instead lead to coral disease and death—the theme of the next chapter.

---

[r] The addition of enormous quantities of nitrogen compounds to the environment by humans is discussed on pages 121-122. Excess nitrogen can impact the coral indirectly by stimulating algal growth and directly by disrupting the in-house nitrogen cycling. One aspect of the latter is its effect on the coral-zooxanthellae symbiosis.

One typically finds approximately one million zooxanthellae per square centimeter of coral surface, give or take some seasonal fluctuations. This equates to 1-2 zooxanthellae per endodermal cell. The coral limits their population by expulsion and also by restricting their available nitrogen to slow their rate of growth. Recall that the holobiont supplies the zooxanthellae with nitrogen in the form of nitrate and ammonium, both of which the zooxanthellae convert into organic nitrogen. The coral polyp then takes back most of the organic nitrogen, leaving the zooxanthellae with but little for their own use. Lacking organic nitrogen, the zooxanthellae can't make more zooxanthellae. It is estimated that each zooxanthellae divides to form two every one hundred days, which is about fifty times less often than their faster-growing, free-living relatives.

When there is a chronic excess of anthropogenic ammonium in the seawater, the coral can no longer limit the amount available to the zooxanthellae, and thus can no longer control their population. In an aquaria experiment where corals were grown in the presence of elevated ammonium levels for 18 days, the zooxanthellae population more than doubled. This might appear to be a good thing: more zooxanthellae, more photosynthesis, more energy for the corals. Not so, because as the number of zooxanthellae increases, each one works less. Not only is there no net increase in the amount of photosynthesis, but also the faster-growing zooxanthellae use more of the photosynthate for their own growth. This all adds up to less sugar and less energy available for the coral, and yet one more way for human activities to stress the holobiont.

Falkowski, P.G., Dubinsky, Z., Muscatine, L., and McCloskey, L. (1993) Population control in symbiotic corals. *BioScience* **43**: 606-611.

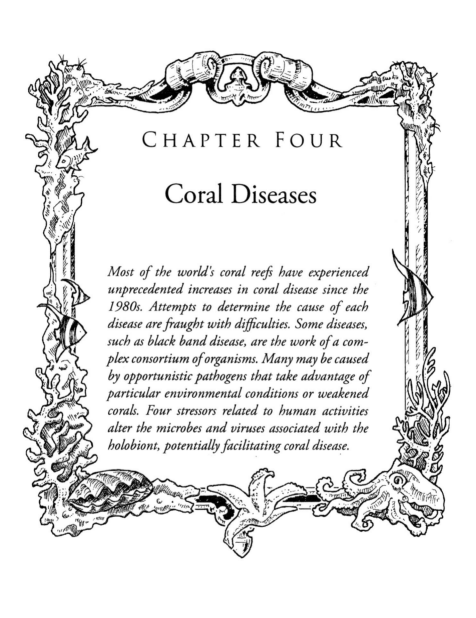

CHAPTER FOUR

Coral Diseases

Most of the world's coral reefs have experienced unprecedented increases in coral disease since the 1980s. Attempts to determine the cause of each disease are fraught with difficulties. Some diseases, such as black band disease, are the work of a complex consortium of organisms. Many may be caused by opportunistic pathogens that take advantage of particular environmental conditions or weakened corals. Four stressors related to human activities alter the microbes and viruses associated with the holobiont, potentially facilitating coral disease.

IV

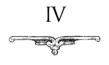

Shark Sampling

With years of experience behind them, the Microbes were extremely good at sampling the viruses and microbes living on corals. They had designed and constructed their own Rube Goldberg sampling devices to slurp said bugs off the docile corals and deliver them into test tubes where they could be properly prodded and probed. This turned out to be kind of boring, as the corals just sat there like slime-covered rocks. This was entirely too tame for one of the Benthics known as the Oldie. She had a much better idea. "Oughta fossick the germs on a flake to see if they're crooked."

"Did anyone bring a Babble Fish, or something else that can translate Aussie into a human language?" inquired a Microbe.

"She's trying to say we should sample the microbes on the sharks. Or she could be asking for another beer. I'm not sure."

"That doesn't sound like a great idea. Sharks bite. Better get her another beer. And you might as well get me one while you're down there."

After a few more beers, the Oldie's shark sampling idea seemed not so bad, and then even a good idea. A Microbe inquired, "How does one sample a shark?"

"We could ask one of the Fish to help. They're professional shark-handlers. And besides, that way, if things go horribly wrong, no humans will get hurt."

For shark sampling the Microbes and the Oldie knew they needed a particular type of Fish. Most people don't realize that there are many distinct species of Fish. There are the Gruff and Grouchy Fish—an older and wiser generation that has survived numerous bad ideas, grumbles a lot, goes to bed early, and generally avoids humanity. Then there are the youthful Minnows, the energetic graduate student Fish, whose exploits will inevitably lead them down the path of Natural (or Unnatural) Selection. They store their safety equipment in their bunk rather than take it out where it might get wet. They rarely bathe. For fun, they consume Tahiti Drink (50% ethanol; 49% methanol; 1% Tang). And they have spent entirely too much time underwater with not enough oxygen. Only one in ten thousand Minnows will survive to grow up to be a Full-Fledged Fish, aka a F3ish.

For sampling sharks, the Microbes needed a F3ish. Somewhat hesitantly, they approached one. "We've got a really dangerous proposal to..."

"Great! Let's do it."

"Don't you want to know what's dangerous about it?"

"Nope, sounds great to me already. When do we start?"

To become a certified shark catcher is the pinnacle of any Fish's career. They spend years mastering the multifarious techniques. Successful completion of the advanced graduate level course is a prerequisite for the hands-on practicum. "First you get a really big hook. Then you attach that to a really strong rope. Next you put a piece of bait on the hook and then you throw it in the water. Make sure to hold onto the rope. I've forgotten that a couple of times...are you guys getting this all down? I might have some notes you could study if you want." The Fish continued to lecture the Microbes. "When the shark takes the hook, you pull him alongside the boat. I'm an expert at this step. Passed this field test during grad school with nine of my fingers still attached."

"What do you do once the shark is next to the boat?"

"Don't worry. That's the fun part." Two hours later, a sun-
burned and dehydrated Microbe peered warily into the water. A
school of vampire snappers had congregated below the boat, hap-
pily eating chum. Frigatebirds circled ominously overhead. These
privateers of the avian world get their name from their practice
of stealing their food, rather than earning it themselves. Normally they harass terns and albatrosses, but boat-bound scientists
are tempting, stationary targets that can be attacked at leisure.
Everyone glanced nervously skyward, and then back to the shark-
less water. More chum. More snappers. More frigatebird attacks.

Then, finally, a shadowy shark-shaped form zipped past.
Then more shadows, and soon even more. Suddenly there were
entirely too many sharks beneath the boat and a feeding frenzy
erupted. Showing the survival instincts of his kind, one of the
Minnows put his mask on and stuck his face into the water. Fins
zipped past. Hastily the Minnow withdrew his slightly lacerated
head and announced in the voice of an expert, *"They're sharks."*

Hooks with bait were thrown into the feeding frenzy. Immediately a F3ish yelled, *"Shark on!"* The F3ish manhandled
the extremely snappy shark over to the gunwale. *"Now just reach
over and grab the tail."* The shark rolled violently, jumped two
feet out of the water, and embedded 50 or so teeth into the side
of the boat. Only then did it occur to the Microbe that an inflat-
able rubber Zodiac might not be the world's best choice for shark
fishing. The slow hiss of escaping air was distinctly audible over
the screeching frigatebirds above and the thrashing sharks below.
"Maybe we should return to the ship and patch the boat?" suggested a slightly shaken Microbe.

*"Don't worry. These boats take forever to sink, even with 50
or more small holes in them. Now someone grab the tail."*

With much splashing and cursing, the Minnow grabbed the
tail. And, of course, the shark went into a frenzy of snapping
teeth. *"Ok, now, one of you needs to grab the shark's pectoral fins
and roll him over so he'll go to sleep."* It suddenly occurred to the
Microbe that the F3ish holding the rope was a safe distance from
the teeth, and the Minnow latched onto the tail was as far from

said teeth as one could be and still be attached to the shark. The pectoral fins were much closer to the wildly snapping maw. Maybe the Fish weren't quite as slow on the uptake as they appeared?!

To make matters worse, Oldie, with her tough Crocodile Hunter accent, turned out to be a total sissy. She conveniently hid as far from the shark as possible, muttering something about needing to have her hands intact for sampling. So the Microbe reached down, grabbed the pectoral fins, and rolled the shark over. "Going to sleep" apparently meant that instead of being dorsal fin up in the water and intent on biting everything, the shark was now upside down and trying to bite everything. In this position, however, the Microbe had a better view of the teeth chomping towards delicate, laboratory fingers.

Once the shark was semi-immobilized, Oldie reappeared with swab in hand and collected a sample of shark-associated microbes. Satisfied, the F3ish announced, "Now all we have to do is turn it loose."

If catching a shark is limb-threatening, turning one loose is downright dangerous—especially if said shark has just been seriously annoyed by being swabbed. "Ok, now, hold onto the pects while I get this hook out. But don't hold him too tight. You don't want to bruise him." At this point, there was only the Microbe at the bitey end and the Minnow at the tail end. "On three! One, two ..." The Microbe let go of the fins on "one" and the Minnow jerked his hands away just in time.

"Well that went great. Let's get another." The Fish were smiling ear-to-ear and giggling. The Microbe and Oldie glanced disbelievingly at each other.

Another baited hook was quickly thrown into the swirling sharkness. A couple baby gray reef sharks in the two foot size range had been observing from the sidelines, leaving the main frenzy to the full-grown five foot grays and white tips. Suddenly one of the babies stupidly darted into the arena and grabbed the hook. Immediately all of the bigger sharks thought, "Oh, here's something easy to eat."

The F3ish was emotionally attached to all sharks. The idea of one getting hurt on his watch was unthinkable. "He's just a baby. Get him in the boat quick, before the others eat him!"

To the Microbe, this seemed to be an exceptionally bad idea. A baby shark already has all of its teeth. Oldie retreated to a distant corner of the boat and silently sobbed. Over the side came the very unhappy, snapping baby gray. Luckily, by now enough air had leaked out of that side of the Zodiac that the hoist wasn't too difficult. The grey thrashed and snapped, the Minnow clutched the tail, the Microbe held the pects, and the ashen-faced Oldie swabbed. Then they sent him overboard, but away from his hungry "friends."

"That was a lot of fun," laughed the Minnow, "and we've still got four more islands where we'll catch some more." Back safely on the main boat, the Microbe and Oldie retreated into beer. "We should stick with the corals. They never bite. And besides, we're never going to find any sharks with diseases. As soon as they get sick and slow down, they're going to get eaten by their acquaintances. What were we thinking?"

iseases are the main killers of corals. Other factors, such as heat-induced bleaching, may weaken the holobiont, but what ultimately kills the coral is disease. Most of the world's coral reefs have experienced unprecedented increases in coral disease since 1980. More than thirty distinct diseases have now been described and more are found every year.

Traditionally, to investigate a new disease, one begins by looking for a *specific* pathogen. This is the microbe or virus capable of causing the disease in a healthy organism. Since the 1880s, the gold standard for the identification of specific pathogens has been Koch's[a] postulates, which state:

1. The microbe must be found in abundance in all organisms suffering from the disease (and preferably not be present in healthy organisms).

2. The microbe must be isolated from the diseased organism and grown in pure culture (i.e., grown from a single cell under laboratory conditions, such as in a nutrient broth or on an agar plate).

3. The cultured microbe must cause the disease when introduced into a healthy organism.

4. The same microbe must be re-isolatable from the newly-diseased experimental organism. (This one was a later addition, not included by Koch himself.)

Unfortunately, it has been difficult to apply Koch's postulates to coral diseases. Black band disease (BBD) provides a case in point. BBD was first described by Arnfried Antonius (Universitaet Wien) in 1973 on the reefs of Belize. It is now known to infect more than forty species of coral around the world. BBD is both highly contagious and susceptible to antibiotics—evidence that it is caused by Bacteria. Warmer water temperatures accelerate its spread, as do extra inorganic nutrients from sew-

---

[a] *Robert Koch* (1843-1910) was a German physician and Nobel Laureate who isolated the bacterial pathogens responsible for cholera, TB, and anthrax. He originated the criteria still used today to establish a causal relationship between a pathogen and a disease.

age and agricultural runoff. BBD is but one of numerous "band" diseases that are characterized by a zone of diseased or dying tissue that moves across a coral colony, advancing as much as several centimeters (an inch) in a day. The band advances onto healthy coral tissue and leaves freshly exposed skeleton and dead corals in its wake.

Although BBD has been intensively studied, Koch's postulates have not been fulfilled. The BBD pathogen will not grow in culture from a single cell—a requirement of Koch's postulates—because "the pathogen" is a pathogenic microbial consortium made up of more than fifty types of Bacteria plus numerous marine fungi.[b] One would have to isolate cells from each member of the consortium and then combine them to reconstruct the complex, interdependent community—extremely difficult, if not impossible, to do.

In these respects, the BBD microbial consortium most closely resembles the microbial mats found in sunlit, sulfide-rich aquatic environments. Although only a millimeter (0.04 inches) thick, the microbial band is multi-layered and highly organized, thereby establishing several distinct microenvironments. The microniches created by the metabolic activities of some member organisms provide suitable habitats for others.

How does the complex BBD mat kill corals? Laurie Richardson and colleagues (Florida International University) used microprobes to measure the local concentration of toxic sulfide in the mat. They found concentrations lethal to corals. This does not, however, mean that the sulfide producer is *the* pathogen, or that sulfide is the only way the mat kills corals. Other mat-associated Bacteria contribute to the high level of microbial activity that consumes all of the local oxygen, leaving the coral hypoxic.[c] This same phenomenon occurs at coral-algal interfaces, as described on page 112. The coral tissue is both poisoned and suffocated by the microbial mat. It takes the entire complex BBD community to cre-

---

[b] Bruce Fouke's group at the University of Illinois and John Bythell's group at Newcastle upon Tyne were the first to strongly suggest that coral diseases might be caused by an ever-changing consortium of pathogens. Their pioneering work employing molecular techniques to study BBD led to the realization that although members of the same bacterial and fungal groups are always present, the particular species that make up the BBD mat are highly variable. A species can be replaced by other closely-related ones that play the same essential role in the community. Among the bacterial members that one always finds are photosynthetic cyanobacteria, sulfide-oxidizing Bacteria (*Beggiatoa* spp.), and sulphate-reducing Bacteria (*Desulfovibrio* spp.).

[c] *Hypoxia* is broadly defined as a condition in animals in which insufficient oxygen is available for body tissues.

ate the lethal microenvironments. *It is becoming clear that many coral diseases are, like BBD, microbial mats that crawl across the coral surface, killing the coral en route by creating hypoxic zones, releasing toxins, and other nasty tricks.*

ℰℐ § ℰℐ

Some of the most important coral diseases are the work of thus-far-unidentified pathogens that kill by as-yet-unknown mechanisms. The most famous of these is white band disease (WBD).

Prior to 1980, the magnificent branching *Acropora* spp.[d] corals dominated the crests of Caribbean reefs. Then, during the 1980s and 1990s, WBD swept through the Caribbean causing almost 100% mortality of acroporid corals and virtually exterminating the staghorn coral, *Acropora cervicornis*. This devastation and loss of coral cover on the shallow Caribbean reefs was unprecedented. A study of deep drill cores from the reefs of Belize showed that there had been no such mass mortalities of this species for at least the past 3,000 – 4,000 years.

Despite its far-reaching impact throughout the Caribbean, the causes of WBD remain elusive. Its advancement is slightly accelerated by warmer waters but is unaffected by antibiotic treatment. The disease is not particularly contagious; experimental inoculation of healthy corals with material from the white band region does not pass on the disease. Attempts to isolate a culturable pathogen from diseased WBD tissue yielded contradictory results. This led Kim Ritchie (Mote Marine Laboratory), Garriet Smith (University of South Carolina), and Ernesto Weil (University of Puerto Rico) to conclude that WBD is really two diseases, now known as WBD type I and type II. A culturable suspect pathogen, the bacterium *Vibrio carchariae*, is characteristically found in the white bands diagnosed as type II, whereas no suspects have been isolated from type I.

WBD type I illustrates some of the difficulties encountered when studying coral diseases. The first hurdle appears when diagnosing diseased

---

[d] The abbreviation *spp.* represents several species within the same genus.

corals in the field.[e] On an ailing coral colony, a sharp line between the bare skeleton and the white band tissue is diagnostic for WBD, whereas a slightly rough or irregular line indicates snail predation. Imagine distinguishing between these when underwater with the surge knocking you around.

Even more challenging than diagnosing a disease is tracking down the cause (or causes). One strategy is to compare the microbial communities associated with healthy and diseased members of the same coral species based on the 16S rDNA sequences present. Microbes found only on diseased colonies are suspect. This was first done with BBD by Bruce Fouke and colleagues (University of Illinois) and John Bythell's group (Newcastle upon Tyne). They found that the BBD microbial mat contained many different types of microbes not normally associated with healthy corals.

When Veronica Casas and colleagues at San Diego State University used 16S rDNA surveys to identify a cause for WBD type I, they found no candidates. However, they did discover that the microbial communities associated with *Acropora* spp.—both healthy corals and those with WBD type I—are dominated by a *Rickettsiales* species that they named CAR1 (coral-associated *Rickettsiales* 1). Although also present on healthy corals, CAR1 is suspect because most *Rickettsiales* spp. are endosymbionts that live inside other cells; many are human and animal pathogens.

This finding, combined with the sudden appearance of white band disease in the 1980s and its rapid destruction of essentially all of the *A. cervicornis* in the Caribbean, led to the hypothesis that WBD type I is a

---

[e] Ailing corals offer us only a few observable symptoms. Coral tissue can die and peel back from the skeleton starting at an edge. This creates a band of diseased or dying tissue. The band can be characterized further by noting the color (e.g., white, black, red). This is how we end up with disease names such as black band disease. In other instances, dead coral tissue can be seen surrounded by apparently healthy tissue. This is called a pox (e.g., white pox disease). Coral tissue can also lift off the skeleton in pieces or all at once, at seemingly random locations, the result of a "wasting disease."

More subtle criteria are often added. Does the disease edge look ragged or smooth? Obviously, this can be difficult to distinguish underwater, and in the end the effort might well be futile. Appearances are often misleading. Coral tissue that looks healthy might already be stressed by an approaching disease. The "same" disease might manifest differently on different coral species. Sometimes, there is even disagreement as to whether an observed condition is actually a disease. Tissue loss or damage from predation can create similar symptoms.

To illustrate the problematic nature of coral reef diagnosis, at the 11th International Coral Reef Symposium (July, 2008), John Bythell and his students circulated an unlabeled photographic rogues gallery of approximately thirty coral diseases. Attendees were asked to diagnose each photo. The results? No one could accurately identify most of the diseases; the experts did no better than novices.

newly emerged coral disease caused by CAR1. The remnant few *A. cervicornis* colonies that survived may represent the small percentage of the population that was naturally resistant to CAR1. This could explain the presence of CAR1 on both healthy and diseased acroporid corals today.

One way to test this hypothesis would be to travel back in time and determine whether or not CAR1 was present on *Acropora* spp. corals prior to 1980. Actually, this could be done to some degree, thanks to the coral collections housed by the Smithsonian Institution. Preserved in alcohol were *A. cervicornis* samples that had been collected between 1937 and 1980. When these samples were screened, no CAR1 was detected, indicating that this potential pathogen was not present prior to 1980. This finding supports—but does not prove—the hypothesis that CAR1 is the cause of WBD type I.

If CAR1 is the cause, then we are left with the question of why the disease suddenly emerged in the 1980s. The answer is unknown and probably unknowable. However, it is notable that an epidemic decimated the Caribbean sea urchin populations in 1983—its cause also unknown. Other diseases suddenly appeared in the Caribbean in the 1980s, as well. Coincidence? Or was there an underlying environmental cause?[f] The answer remains a mystery.

ༀ § ༀ

---

[f] One hypothesis put forward for the abrupt increase in disease in the Caribbean is that sudden global climate shifts brought greater amounts of iron to the reefs in that region. Iron is a key, growth-limiting nutrient in the oceans. Competition for iron also plays a role in many pathogen-host interactions, with both sides seeking to lay claim to the needed resource. A sudden infusion of iron into Caribbean waters could have removed the limitations that normally keep opportunistic and potentially pathogenic microbes in check.

Atmospheric dust is the major source of new micronutrients, including iron, for the ocean's surface waters. The amount of dust in the atmosphere fluctuates in response to numerous climatic and anthropogenic factors. The global climate shift that occurred in the mid-1970s ushered in a prolonged and historically unprecedented drought in regions of Africa and also heightened easterly trade winds over the Atlantic. Combined, they brought a dramatic increase in windblown dust to the Caribbean region, an increase continuing to today.

Another story linking climate shifts, increased iron, and coral death is provided by the detailed observations made in Indonesia following the 1997 algal bloom, or "red tide." These regional red tides are brought on by aperiodic climate shifts that increase the upwelling of nutrients to the coral reefs around Sumatra, Indonesia. The upwelling that occurred at the end of 1997 was not unusual, but its effects were extraordinary. The accompanying red tide caused catastrophic coral death at a scale unprecedented for at least 7000 years. Again the question: why then? That was also the time of the worst wildfires in the recorded history of southeast Asia. The data suggests that this red tide was enhanced by extra iron from the ashfall produced by the nearby fires. Although only a hypothesis, the pieces of the puzzle fit together neatly.

Although it is generally agreed that most of the world's coral reefs have experienced unprecedented increases in coral disease since 1980, no one knows exactly why. Both the global and the local effects of human activities are highly suspect. Leading the list of potential disease-facilitating stressors is the rising sea surface temperature. Temperature-induced coral bleaching makes corals more susceptible to opportunistic pathogens, and some pathogens are more virulent at higher temperatures. So far, five known coral pathogens have been found to be most virulent when the seawater temperature is 29 °C or higher. Coral defenses may also weaken with elevated temperature. Currently, we know very little about the possible role of increasing acidity on the incidence of coral disease. The local stressors, nutrient enrichment and overfishing, are usually associated with increases in coral disease. The situation is complicated, and there are likely other, not-yet-identified stressors that act in concert with these four.

The complex interplay of factors that can combine to produce an observed coral disease is elegantly portrayed in the microbial bleaching story that was uncovered by the Israeli trio of Eugene Rosenberg (Tel Aviv University), Ariel Kushmaro (Ben-Gurion University of the Negev), and Yossi Loya (Tel Aviv University). In 1993, Loya observed that about 80% of the *Oculina patagonica*[g] corals along Israel's Mediterranean coast bleach each summer. Most of them recover during the cooler winter months. Loya teamed up with Rosenberg, a distinguished microbiologist, and together they put Kushmaro, a PhD student, to work to research the cause.

Kushmaro's research uncovered a fascinating story. He demonstrated that bleaching of this coral was due to a specific bacterial pathogen, *Vibrio shiloi*.[h] Moreover, the pathogen infected and bleached the corals only in the warm (~29 °C) waters of summer. In winter water temperatures (16-20 °C), it was unable to synthesize three virulence factors needed for successful infection and bleaching. The first factor is an adhesin, a

---

[g] To date *Oculina patagonica* is the only coral known to have migrated globally, in this case from the southwest Atlantic Ocean to the Mediterranean Sea. It arrived there in 1966, presumably as a hitchhiker attached to the hulls of ships plying that route. Also atypical is its ability to tolerate greater variation of both temperature and salinity.

[h] Rosenberg, E., and Falkovitz, L. (2004) The *Vibrio shiloi* / *Oculina patagonica* model system of coral bleaching. *Annual Review of Microbiology* **58**: 143-159.

protein used by the pathogen to attach to the coral. The second is a toxin that disrupts photosynthesis by the zooxanthellae, leading to their expulsion and thus the bleaching of the coral. The third is a defensive enzyme needed by the pathogen to break down the reactive oxygen species (ROS) produced by the zooxanthellae during photosynthesis. Without this protection, the vibrios die from ROS-inflicted damage to their DNA.

This explained why the corals were bleaching only during the summer, but it begged the question of where *V. shiloi* overwinters. It turns out that it can be found year-round on the reefs in the gut of the bearded fireworm *Hermodice carunculata*, a voracious predator of corals. The fireworms serve as a reservoir for the pathogen, continually infecting the corals as they graze upon them. However, it is only in the summer months that *V. shiloi* can produce the factors it needs in order to attach, survive, and disrupt the coral-zooxanthellae symbiosis.[i]

In their follow-up work, Rosenberg and colleagues established that *O. patagonica* is not the only coral bleached by microbial pathogens. Their second case is the coral *Pocillopora damicornis* that bleaches when infected by *Vibrio coralliilyticus* at elevated temperatures. Neither *V. shiloi* nor *V. coralliilyticus* can breach the defenses of healthy hosts under normal conditions. They may even be normal members of the healthy holobiont. They can cause disease only when elevated temperatures increase their virulence and/or stress the corals.

These two prototypes of vibrio-induced bleaching illustrate how intimately interwoven are the microbial and environmental causes of coral disease. There is yet a third major interacting factor: the diverse defense strategies employed by the coral. These are discussed in Appendix B.

ᘓ § ᘓ

---

[i] This made for a very satisfying story. From 1995 until 2002, *V. shiloi* was always present on the summer-bleached *O. patagonica* on the reefs. When healthy corals were collected and infected in the lab with *V. shiloi*, the corals bleached. Then, in 2003, *V. shiloi* unexpectedly disappeared from the reefs. Furthermore, laboratory stocks of *V. shiloi* collected pre-2003 that had been able to infect *O. patagonica* corals could no longer bleach newly-collected corals. The corals had become resistant to *V. shiloi* infection by some unknown mechanism. The researchers suggested one possible means: the corals might now host a population of some other Bacteria that blocks infection by *V. shiloi*.

Currently, we do not know what mechanism is in play, and the not knowing leaves plenty of room for spirited debate. Nevertheless, this observation of a recently-acquired disease resistance is heartening evidence that the coral holobiont has adaptive maneuvers that we have yet to discover.

While more than thirty coral diseases have been described so far, Koch's postulates have been fulfilled for only eight[j] —and even for these, there are variable degrees of certainty. *This suggests that many coral diseases may not be caused by specific pathogens, but rather are the work of opportunistic pathogens in league with environmental stressors.* The investigation of opportunistic diseases that require the participation of environmental factors calls for a new strategy, one that enables us to monitor the interaction between environmental stressors and the holobiont's microbial community. Becky Thurber (San Diego State University) showed that metagenomics is an effective approach for studying the effects on the holobiont of four known environmental stressors: increased seawater temperature, decreased pH, increased inorganic nutrients, and increased dissolved organic carbon. The

Becky Thurber

first two are by-products of increased atmospheric CO_2 and thus can affect corals in any region. Nutrient addition is a local consequence of human activities such as agriculture and sewage dumping. Increased dissolved organic carbon is associated with overfishing of the reefs and is a local stressor of such importance that we will spend much of the rest of the book discussing it.

In Thurber's experiments, fragments ("nubbins") of the branching coral *Porites compressa* were placed in aquaria where they were subjected to one of these four stressors for two and a half days. Afterwards, the

---

[j] Most of those eight are the work of the team of Kim Ritchie, Garriet Smith, and Ernesto Weil. They were the first to fulfill Koch's postulates for a coral disease (white pox). Ritchie contributes the expertise in both microbiology and molecular techniques needed to identify the potential pathogen. Her approach is complemented by that of Ernesto Weil, an extremely hardworking field biologist with a knack for identifying new diseases. He has documented numerous diseases around the world, most notably in the Caribbean. The elder statesman of the group is the eccentric and fun-loving microbiologist, Garriet Smith. Together, this trio inspired a new generation of microbiologists to enter the challenging field of coral disease.

microbes associated with the nubbins were collected, and then their DNA was extracted and sequenced. When the resulting microbiomes were analyzed, it was found that all four treatments altered the microbial community associated with the coral. In all cases, the community shifted from that typically associated with healthy corals to one dominated by the microbes found with diseased corals. Likewise, in all four cases the microbiomes associated with stressed corals contained more genes for virulence factors, suggesting that more pathogens were present.

Despite these shared trends, the specific changes were distinctive for each stressor. For example, elevated water temperature increased the number of vibrios associated with the coral. This result would have been considered significant simply because vibrios include numerous pathogens. It was particularly striking because of the previously mentioned role of vibrios in coral bleaching (see pages 80-81).

We all know that many of our illnesses are caused by viruses, not Bacteria. Probably every species on the planet is host to some virus. Even Bacteria have their phages.[k] It is extremely likely that viruses cause some coral diseases. However, at present the viruses associated with corals represent the largest *terra incognita* within the holobiont community. Early reports implicating viruses in coral disease have come from Nicole Patten in Australia (Southern Cross University) who found virus-like particles associated with disease lesions in white patch syndrome, and from Willy Wilson's group in the UK (Plymouth Marine Laboratory) that linked viral infection of zooxanthellae with coral bleaching. Although provocative, follow-up investigations have been few because the traditional techniques for studying viruses require propagating them in the lab. That, in turn, requires being able to culture host cells in which the viruses can reproduce. So far, no one has developed a procedure for maintaining coral cells in tissue culture.[l] So researchers are, of necessity, turning to metagenomics and other culture-independent techniques.

Kristen Marhaver (Scripps Institution of Oceanography) was the first

---

[k] A *bacteriophage*, or *phage*, is a virus that infects Bacteria.

[l] *Tissue culture* is the growing of cells or tissues separate from the source organism, typically in a sterile growth medium. Cell lines can be maintained long term, through many generations, and used to study the physiology of the organism, including the changes associated with infection and disease.

to use metagenomics to characterize the viruses associated with a coral. Her investigation of the virome[m] of the Caribbean reef-building coral *Diploria strigosa* netted stunning results. *There are 1,000,000,000,000 (10^{12}) viruses per square centimeter of coral surface. The coral-associated viruses are also astonishingly diverse. The virome from* Diploria strigosa *contains an estimated 28,600 different viral species.*

This highly diverse virome contains viruses that could potentially infect every member of the holobiont, including the coral, zooxanthellae, Bacteria, and Archaea. Of the viruses that might infect the coral animal, a large majority are herpes-like.[n] The potential significance of these herpes-like viruses was documented by Thurber when she monitored the effects of stressors on the viromes of her experimentally stressed corals (see above).[o] *Elevated temperature, nutrient additions, and lowered pH all dramatically increase the number of herpes-like viruses associated with the corals.* In a series of follow-up studies, Thurber found that the increase in viruses was very quick, within 4 to 6 hours. This strongly suggests that herpes-like viruses are residing in the holobiont and are reactivated by stress—a situation analogous to our latent herpes infections that, when reactivated, cause cold sores.[p]

Much work remains to be done to elucidate the role of environmental stress in coral disease and death. In particular, we need to understand how coral diseases are assisted by the two major local stressors: overfishing and nutrient enrichment. Their stories unfold in the next three chapters where we will see that these stressors act by benefiting the algae and the microbes.

---

[m] A *virome* is a metagenome made up of viral genomes.

[n] Finding so many herpes-like viruses in the coral virome was unexpected. For a long time, herpes viruses had been thought to infect only vertebrates; then a subgroup was identified that infects oysters and other bivalve molluscs raised in aquaculture. Now corals have been added to the list. Intriguingly, herpes-like viruses usually infect neurons, and Cnidarians were the first animals to have a nervous system. A second surprise came when these herpes-like viruses were found to be more closely related to those infecting humans and other vertebrates than to those that infect other invertebrates.

[o] Thurber, R.V., Willner-Hall, D., Rodriguez-Mueller, B., Desnues, C., Edwards, R.A., Angly, F. et al. (2009) Metagenomic analysis of stressed coral holobionts. *Environmental Microbiology* 11: 2148-2163.

[p] In people, after the initial active herpes infection, latent viruses often reside for long periods inside nerve cells. When reactivated by illness or stress, they migrate to the skin where they replicate and produce the symptoms.

CHAPTER FIVE

Overfishing and the Rise of the Algae

Coral reefs are complex ecosystems composed of many interacting variables. Their mind-boggling complexity can be visualized as a biogeochemical space. In the 1980s, the reefs around Jamaica quickly changed from a coral wonderland to an algae-dominated landscape. Similar shifts have now been observed on reefs around the world. These shifts to an alternate reef state characterized by lots of algae correlate with local overfishing. This leaves us with the question, how does the greater abundance of algae kill the corals?

V

Fishing for Dinner

A person burns approximately 2000 calories on a typical day. SCUBA diving adds another 400 calories per hour. So an average diver, making three one-hour dives a day, needs over 3000 calories. On the Northern Line Islands Expedition, there were twenty people on board, each making daily dives for five weeks. Doing the math, one arrives at something like 2.1 million calories needed for the trip, or roughly 1500 pounds of food. A total of 500 pounds of food was loaded, along with 1000 pounds of turkey sausage. When the trip was over, 995 pounds of turkey sausage remained, and there was a net weight loss of twenty pounds per person. The Turkey Sausage Miracle Diet.

The expedition began with the assumption that the Fish would be prepared to…well…fish. However, due to their complex packing list of clipboards, pencils, and defective underwater paper, they forgot to include fishing tackle. In fact, everyone involved in calculating the necessary amount of food had assumed that fishing would proceed naturally with the fish doing their part by jumping onto the boat.

Luckily, unlike the Fish who counted fish all day in the cool, azure waters, the Microbes toiled in the hundred-degree plus heat of the on-board lab and became very bitter toward the Fish and toward the fish with their carefree, free-swimming, sashimi-avoiding ways. A plot to catch and eat fish was hatched and the necessary gear was jury-rigged from assorted junk scavenged from the several tons of microbial field gear.

The first fishing trip took place off Fanning Island. Near the entrance to the lagoon on the leeward side of the atoll is an area where the coral falls off sharply, forming a steep wall. Here the tuna congregate to hunt among the schools of bait fish, picking off the stragglers. A boat passing overhead looks like a bait ball to the tuna eye. This makes them an easy catch for any human. The Microbes set out in one of the dive boats, dragging a lure behind. Wham! A tuna hooked itself and sashimi was on the menu.

Upon their triumphant return to the White Holly, the Microbes turned their soon-to-be sashimi over to the Chef who tied the tuna to one of the small cranes by its tail, hoisted it out over the water, and cut its throat to drain the blood. Predictably, the rope broke and the prized tuna protein slid into the sea. Someone dared to ask, "What does turkey sausage with wasabi and rice taste like?" The Fish, however, sprang into action. Here was a game they understood: FIND THE FISH. It was dark, there was blood in the water, and the Fish had a professional interest in sharks. There was really only one choice. Position a Microbe on the surface as bait to distract the sharks long enough for a Fish to dive and retrieve the sashimi.

A Fish slid into the water with flashlight in hand, the Microbial bait bobbing at the surface. The Fish dove. Immediately there came much screaming from the beer-swilling spectators still aboard. "Sea snake! Watch out for the sea snake!"

There, mere inches from the Microbe's mask, was one of the most deadly creatures on the planet. The snake was somewhere in the five foot range with inch-long fangs that dripped venom. Much to everyone's surprise, the Microbe was prepared for such an occasion. Unlike many landlubbers, he knew that sea snakes are extremely frightened by high-pitched screams and water splashed in their eyes. The Microbe promptly released a 125 decibel, high C to stun the vicious sea serpent, followed immediately by a sound thrashing of the water. With the Microbe thus nobly protecting his back, the Fish retrieved the tuna from the bottom and brought it back on board. The sea snake, still circling in the light, deflated to a mere eight inches and transformed itself into a striped eel— probably due to the Microbe's quick defensive maneuvers.

The fresh sashimi was a great addition to the spartan diet aboard the Holly. Unfortunately, seafood is so tasty and so easy to catch that we humans are quite literally killing most of the fish in the ocean.

n the mid-1980s, most of the coral reefs surrounding the island of Jamaica died. Prior to this collapse, tourists and coral reef ecologists alike had been drawn to these colorful reefs with their abundant and varied corals. Much of the seminal work in coral reef science had been done at Discovery Bay on Jamaica's north coast.[a] Here corals covered 30% to 60% of the bottom, while fleshy algae[b] occupied less than 5%. The collapse of the reefs in the 1980s flipped these percentages, with coral cover falling to 5% and algal cover increasing to about 70%. Since then, similar rapid transitions from a coral wonderland to algal slime have been reported on numerous reefs around the world. Invariably they bring suffering to the local communities as the reef fisheries collapse and tourism plummets. To understand what brings about these *phase shifts* from one state to another, we need to think of a coral reef as an ecosystem.

Coral reefs are more than an assembly of coral colonies; they include thousands of different species interacting with one another in exasperatingly complex ways. Adding another layer of complication, these organisms not only depend on their surrounding environment, they also modify it. The number of possible interactions is so overwhelming that, in order to see the big picture, we have to condense the complexity into something we can visualize.

To illustrate, let's start by asking why corals build reefs in certain locations and not in others. An easy first step is to mark the locations of all known coral reefs on a globe. This physical map doesn't reveal all the reasons why the reefs are located where they are, but by looking at their locations you can surmise some possible factors. For example, such a map clearly shows that coral reefs are found overwhelmingly in the tropics (i.e., between the latitudes 23.4° S and 23.4° N). This zone coincides with water temperatures between 18 °C and 30 °C, and experiences high-

---

[a] Discovery Bay Marine Laboratory in Jamaica was founded in 1965 and quickly became a major center for coral reef research. It was at Discovery Bay in the 1960s and 1970s that the classical architecture of a Caribbean reef was described by Thomas F. Goreau and others. Each zone was characterized by the presence of particular coral species. For example, on the seaward face that takes the brunt of the wave action one found massive fences of living elkhorn coral, *Acropora palmata*. By the early 1990s, most of those *Acropora* thickets and the other coral topologies described by Goreau were no more. Acutely aware of the loss, Goreau's son, Thomas J., is an active coral conservationist.

[b] *Fleshy algae* refers to the seaweeds and the turf algae, both of which are grazed by fish and other herbivores. Both can grow quite large when the number of grazers has been reduced by fishing, disease, etc. For a more complete description, see pages 101-104.

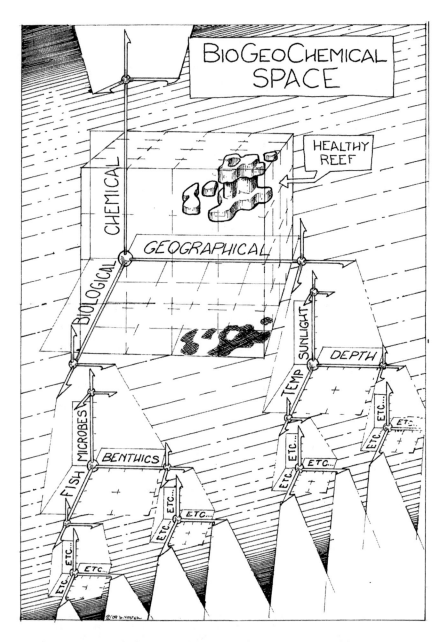

Figure V-1: Many different combinations of interacting variables can support a healthy coral reef ecosystem. The numerous biological variables are represented in our graph by a single proxy that is plotted along one axis, and likewise for the geological and chemical variables. The cascading axes are an attempt to represent a hyperdimensional space on a two-dimensional sheet of paper.

intensity sunlight year-round. This, in turn, suggests that temperature and/or the amount of sunlight may be important when identifying suitable coral habitat.

A closer look at the globe reveals that coral reefs are not found near the mouth of large rivers. The Amazon, for example, separates the reefs along the Atlantic coast of South America into two groups, the Brazilian and the Caribbean. This observation suggests that corals, adapted to normal seawater, do not thrive in less saline waters—a hypothesis corroborated by the coral death observed when extremely heavy rainstorms flood reefs with freshwater runoff from the nearby land. However, other interpretations are possible. Perhaps the sediment discharged at the river mouth smothers the corals or clouds the water enough to reduce photosynthesis by the zooxanthellae. Most disconcertingly, you can find evidence to support all of these interpretations. So we're left with the knotty question: which *variables*[c] are the most important for determining where coral reefs thrive?

In reality, there are thousands of interacting variables that combine to make a location favorable or unfavorable for reef-building corals. It's impossible to mentally envision all of them. Instead, we resort to a method that compresses all of this complexity into three composite variables that can be displayed on a three-dimensional graph (Fig. V-1). The traditional three axes on this graph represent the biological, geographical, and chemical variables. Together they define the *biogeochemical* space occupied by coral reef ecosystems.

Although the variables represented by each of the three axes can have a wide range of values, coral reefs thrive only within a narrow portion of that range. In our sample graph, latitude-longitude is plotted on the X-axis. This one variable serves as a proxy for thousands of geographical features including temperature, day length, upwelling, currents, and coastal topography. Since many of these variables are interdependent, they are said to *covary*. For example, both the amount of sunlight and the temperature will increase as you move towards the equator. The Y-axis represents biology. Our favored proxy here is the abundance of sharks,

---

[c] A *variable* is a quantity that can assume different numerical values, such as seawater temperature, salinity, or pH.

an indicator not only of the populations of large predator fish, but also of the numbers of microbes, viruses, corals, and other reef organisms. By plotting shark abundance, we collapse a large number of biological variables into a single dimension. The Z-axis on our graph represents the chemistry of the seawater. As proxy, we might plot the concentration of nutrients—an important variable since coral reefs thrive in nutrient-poor waters. Let's see how this approach (explained further in Appendix C) might help us understand what happened to the Jamaican reefs.

ᘒ § ᘒ

The collapse of the Jamaican reefs in the 1980s provided a unique opportunity to scrutinize the events that lead to an ecological disaster. As a first step, suppose we use our three-dimensional biogeochemical space to plot all the data measured on those reefs at different times prior to the collapse. We would then see that there are slight differences in the values of the measured variables between one time point and another. These differences may be random variations, experimental error, or real alterations that reflect environmental fluctuations such as seasonal shifts. Thus, the plot for each time point would be slightly different, but every instance would still correspond to a coral reef. Even the same coral reef may not have exactly the same biology, geography, and chemistry at all times. Plotting the data from all available time points for all coral reefs defines a three-dimensional volume on the graph that corresponds to coral reef ecosystems. In other words, there are many different combinations of variables that define functionally-equivalent ecosystems, in this case different instances of the coral reef ecosystem. Together, they define the ecosystem as a *stable state*.

Any ecosystem can exist in any of two or more discrete stable states. To visualize these states, picture a mathematical landscape with two or more valleys. Each valley corresponds to a possible stable state of the ecosystem. Now imagine a ball in one of the valleys. This indicates the current state of the ecosystem. In geek speak, each valley represents the basin of attraction for a particular ecosystem state. When lightly "pushed," the ball will roll around in the valley. These pushes represent changes in the environment and are called *perturbations*. Most perturbations will not knock the ball out of its current valley; the ball simply rolls back downhill to the valley floor. However, a strong perturbation can drive the ball from one valley to another, i.e., drive the ecosystem from one state to another. Once the ball crosses over the dividing ridge, it will continue rolling downhill to the bottom of the new valley. The ecosystem is now in a different stable state.[d] When looking at coral reefs, what we ultimately want to do is identify the perturbations that drive their transition from having lots of healthy corals to another state, such as one where the algae dominate the reef.

This model also enables us to visualize some of the factors that make an ecosystem resilient.[e] The size of the valley indicates the amount of change that an ecosystem can absorb without shifting to a different stable state—a property that is referred to as *latitude*. The depth and shape of the valley also reflect the *resistance* that must be overcome in order to change to another state. The more resistant an ecosystem, the harder it is to move the ball, and thus the stronger the perturbation required to propel the ball out of that valley. A resilient ecosystem can deal with even large perturbations without shifting to an alternative state. Similarly, the steepness of the valley walls represents how quickly the system can recover from a perturbation. When the slope is gradual, the ball rolls more slowly and it takes longer for it to return to the valley floor. Until it arrives there, even a small perturbation might push the ball the rest of the way out of the valley. An example of this would be a coral reef that is less resilient while it is recovering from a major hurricane.

When we apply this sort of analysis to Jamaica, we can identify three valleys in the landscape, each representing a different stable state. One valley represents the post-collapse Jamaican reefs of today. Since they are characterized by few fish and lots of large, fleshy algae growing on the

---

[d] Transitions from one stable state to another often exhibit a curious property, the name for which is *hysteresis*. Sometimes the current state of a system is determined not only by the current conditions, but also by past history, i.e., by how it got to where it is now. Remember the straw that broke the camel's back? Removing that straw—i.e., restoring the original conditions—does not restore the camel to its original state. That's hysteresis. When applied to ecosystems, it warns us that although it may be easy to push an ecosystem into a different state, it may prove far more difficult to restore it to its original one.

[e] The *resilience* of an ecosystem is the ability of that ecosystem to return to its original state following a perturbation.

skeletons of dead corals, we'll call this the *algal state*. The second valley, which we'll call the *urchin state*, is characterized by low numbers of fish, lots of coral, little algae, and enormous numbers of the long-spined, black sea urchin *Diadema antillarum*. Available research data shows that this state characterized the same coral reefs during the pre-collapse years from 1950 to the mid-1980s. The third stable state, the *fish state*, represents the reefs before the 1700s. Based on historical records and coral coring, we know that the reefs had high coral cover and abundant populations of large fish and large marine mammals.

What were the perturbations that pushed the precolonial reefs into the urchin state, and from there into the algal state? The primary perturbation was fishing. In the years since the arrival of Columbus in 1494, Caribbean fish stocks have declined by 80%. Most drastically reduced are the populations of sharks, snappers, groupers, and other large fish predators, along with other large vertebrates such as turtles, manatees, and monk seals. By the 1960s, with the more desirable larger species scarce or completely gone, fishermen had resorted to catching the smaller, herbivorous species that grazed the algae on the reefs. Times were good for the sea urchin *Diadema*. With fewer predator fish eating it and fewer algae-grazing fish competing with it, *Diadema* flourished. Its population reached more than ten hungry urchins per square meter (yard)—enough mouths to keep the fleshy algae in check.

The collapse from the urchin to algal state occurred in 1983 when a devastating epidemic struck the sea urchins, rapidly killing more than 99% of them across the Caribbean. The pathogen responsible was never identified. The few remaining herbivorous fish could not graze down the fleshy algae, and the algae killed the corals. At this point, the reefs crossed the divide from the urchin state to the algal state.[f] Jamaica was left with a

---

[f] In reality, the story is not so simple or tidy, as there were other contributing factors. First, Discovery Bay had been hit by Hurricane Allen in 1980, the strongest hurricane recorded in the Caribbean up to that time. The coral death toll continued to rise for months afterwards. Then, in 1987, Discovery Bay suffered one of the first recorded large-scale coral bleaching events brought on by abnormally high sea surface temperatures. Abundant large fleshy algae, a result of the reduced sea urchin population, covered the bottom, thus preventing coral recolonization. There may have always been relatively high numbers of *Diadema* urchins on Caribbean reefs, and they may even have gone through similar die-offs in the past. However, without the fish grazers, the algae grew unchecked following the 1983 epidemic.

Hughes, T.P. (1994) Catastrophes, phase shifts, and large-scale degradation of a Caribbean coral reef. *Science* **265**: 1547-1551.

decimated fishery and—without the beautiful coral gardens—a foundering tourist industry.[g]

This scenario of overfishing leading to increased fleshy algae and then coral death has been observed time and time again on coral reefs around the world. We are left with a question: how do algae kill corals?

---

[g] Can the Jamaican reefs recover? Coral recovery will require, among other things, reduction of the algal cover to create open space where young corals can settle to recolonize the reef. Since intense fishing still keeps the population of algae-grazing fish low, there aren't enough fish to graze down the algae. However, the sea urchins are making a comeback. A rebounding sea urchin population might drive the shift from the algal state back to the urchin state. *Diadema* has become locally abundant again in the shallow waters of the fore reef at Discovery Bay. Already the urchins have reduced the percentage of substrate covered by fleshy algae, and in those areas one now sees increased recolonization by numerous key coral species. Of particular importance to the community are the sturdy, hurricane-resistant *Montastraea* corals, the frame-builders of the Caribbean reefs. These long-lived corals colonize and grow slowly; their recovery time, at best, will be measured in decades.

Although the comeback of the sea urchins is so far only a local development, it nevertheless gives us hope for the recovery of the devastated Caribbean reefs as a whole. Our optimism, however, is tempered by the disproportionate importance of this sea urchin for reef recovery and long-term stability. So long as the reefs remain dependent on a single herbivore, they will remain vulnerable to a repeat of the 1983-84 collapse. Having robust populations of more than one organism grazing the algae would make the reef more resilient in the face of storms, diseases, or other perturbations. Thus, reduced fishing to allow recovery of the herbivorous fish populations is essential for the long term health of these reefs.

Carpenter, R. C. and P. J. Edmunds (2006). Local and regional scale recovery of *Diadema* promotes recruitment of scleractinian corals. *Ecology Letters* **9**: 271-280.

Idjadi, J. A., S. C. Lee, et al. (2006). Rapid phase-shift reversal on a Jamaican coral reef. *Coral Reefs* **25**: 209-211.

Knowlton, N. (2001) Sea urchin recovery from mass mortality: new hope for Caribbean coral reefs? *Proceedings of the National Academy of Sciences USA* **98**: 4822-4824.

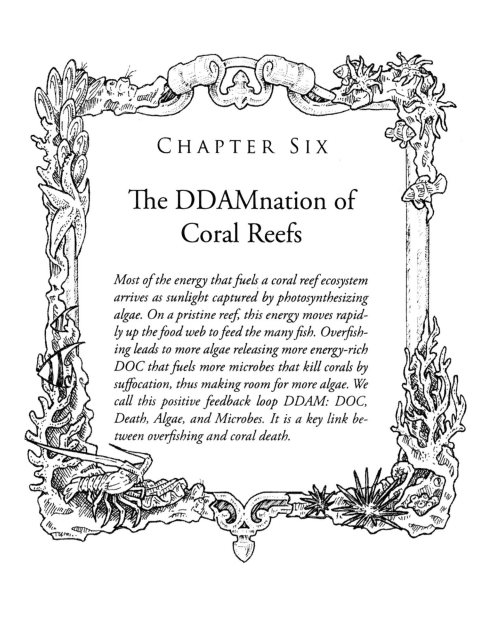

CHAPTER SIX

The DDAMnation of Coral Reefs

Most of the energy that fuels a coral reef ecosystem arrives as sunlight captured by photosynthesizing algae. On a pristine reef, this energy moves rapidly up the food web to feed the many fish. Overfishing leads to more algae releasing more energy-rich DOC that fuels more microbes that kill corals by suffocation, thus making room for more algae. We call this positive feedback loop DDAM: DOC, Death, Algae, and Microbes. It is a key link between overfishing and coral death.

VI

The Bitch

Long before the Northern Line Islands Expedition left the dock, an earnest grad student, whom we'll call the Worker, had set out to investigate how fleshy algae and other environmental stressors might be killing corals. To do this, he needed to incubate hundreds of coral fragments, or nubbins, with different stressors. But no such culturing system existed. So the Worker recruited the Tweaker, a meth-powered genius who built things, then took them apart, rebuilt them a little "better," took them apart again, ad infinitum. To break the cycle and extend the life span of some of the products, the Worker settled onto the Tweaker's couch where he could nab newly constructed modules before they were disassembled. Eventually the Worker emerged with the Aquatic Automated Dosing and Maintenance System, or AADAMS, destined to be known to all that worked with her as The Bitch.

The Worker and The Bitch were loaded onto an airplane and shipped to Panama. There, progress was slow. The Bitch was calling the shots. And so, an ad was posted.

Wanted: World class mechanic who enjoys tropical weather and biting insects. Must love maintaining recalcitrant machinery without supplies or proper tools. Salary zero or some multiple thereof, depending on experience.

Surprisingly, the ad did not elicit a wave of wannabe employees. In fact, only one person responded, an itinerant known as the Misdirected. The Misdirected's first job in Panama had been as a fish shocker. This is an individual who wanders about electrocuting fish with a steel prod connected to a car battery. To avoid killing oneself with a bolt of lightning, this is done in heavy-duty rubber waders while toting the fifty pounds of battery through the 82 °F temperature and 100% humidity of the deep rainforests. Anyone foolish enough to be a fish shocker was obviously perfect fodder for The Bitch.

Together, the Worker and the Misdirected cut the tips off hundreds of coral colonies with chisels and hammers. These precious nubbins went into the incubation chambers watched over by The Bitch where they were each awash in one stressor or another. During every experiment, The Bitch invented novel and sadistic ways to not work, while the sand fleas grew fat on the blood of the Worker and the Misdirected.

Despite their slow torture, most of the coral nubbins survived for the duration of the experiment, except for those treated with extra dissolved organic carbon (DOC). This small piece of evidence turned out to be the crucial link connecting overfishing with the death of corals on reefs around the world.

ost of the energy that fuels coral reefs arrives as sunlight.[a] *Primary producers* are the organisms that use the sun's energy to synthesize sugars and other organic compounds from CO_2 and water. The overall process is called *photosynthesis*; the specific steps that capture CO_2 are termed *carbon fixation*. Primary producers are also called *autotrophs* ("self-feeding") because they can feed themselves. They also feed all of the *heterotrophs* that need their energy supplied in the form of organic compounds made by the autotrophs.

The principal autotrophs on reefs are cyanobacteria and several types of algae. The algae span a size range from the zooxanthellae that live inside coral cells to the larger reef algae that are grazed by parrotfish, sea urchins, and other herbivores. These herbivores are, in turn, eaten by predators that are eaten by yet larger predators. This who-eats-whom makes up the *food web*, and it can be quite complicated.

Trophic pyramids are one way to simplify our view of food web complexity. These imaginary pyramids are built in layers, each layer being a *trophic level* made up of all of the organisms that occupy similar positions in the food web. All the organisms on one level eat those on the level below them and are eaten by those on the next level up. At the base of the pyramid are the primary producers, the first trophic level. The grazers that feed on them make up the second trophic level; the predators that consume them make up the third; and so on up the pyramid to the top, or apex, predators in the community. Sharks, of course, are the most famous apex predators on coral reefs.

One can construct trophic pyramids that reflect the number of organisms at each level, or the total biomass in the organisms at each level, or—most telling—the amount of energy residing in organisms at each level. The energy at the first trophic level is derived from the sunlight harvested by the autotrophs and stored as chemical bonds. This energy moves up the pyramid through grazing and predation. It is used by the organisms at each trophic level to build their biomass and fuel all their ac-

---

[a] Some of the energy and nutrients drift or swim in, instead. The incoming water is laden with usable food energy in various forms. The front of a coral reef facing the open ocean—the "wall of mouths"—functions as a massive and extremely efficient filtering device. Small fish, corals, and crustaceans snatch up incoming particles of plankton, while sponges capture the DOC, viruses, and microbes. The organisms that make up the wall-of-mouths excrete their nutrient-rich waste products onto the reefs where they are accessible to other members of the community.

tivities— everything from metabolizing their food to swimming around on the reef. *The rough rule of thumb is that of the energy "eaten" or captured by an organism, about 90% is spent for its own activities and metabolism. Only about 10% is passed on to the next higher trophic level.*[b] As a result, there is always more energy at the base of the pyramid, successively less at each level above the first.

This rule helps us to estimate just how much algae it takes to make a shark. A reef shark weighs about 45 kilograms (100 pounds). A 45 kilogram shark had to eat 450 kilograms of parrotfish that, in turn, had eaten 4,500 kilograms (five short tons!) of algae. Because reefs actually have more than three trophic levels in their pyramids, there are usually more layers taking their 90% cut of the energy as it makes its way to the sharks at the top. As a result, these figures are underestimates. The take home point is that each shark represents an enormous amount of sunlight and algae.

⁋ § ⁋

There are many types of algae on a coral reef. The zooxanthellae that live and photosynthesize within the coral polyps are single-celled algae that make up a group within the dinoflagellates. Very little of the sun energy captured by them makes its way to a shark's tummy, and then only by circuitous routes.[c]

In contrast, other algae feed sharks more directly. The most important are the diminutive *turf algae* that make up the lawns on the reef surface. "Turfs" are consortia of small algae, typically comprising more than 20 species and sometimes 200 or more. They are exceptionally nutritious, thanks to the high protein content contributed by the nitrogen-fixing cyanobacteria that live in the lawns. These tasty turfs are heavily grazed

---

[b] This rule of thumb is more accurate for endotherms (the "warm-blooded" birds and mammals) and less accurate for the "cold-blooded" ectotherms, such as fish and marine invertebrates. Since ectotherms regulate their body temperature mainly by exchanging heat with their environment, they convert less of their food energy into heat. As a result, they use less than 90% themselves and thus more than 10% is passed on up to the next trophic level.

[c] Between 20% and 45% of the photosynthate made by the zooxanthellae is devoted to mucus production by the coral. The surface mucus layer is eventually shed by the coral, then consumed by plankton-eating fish and reef-dwelling invertebrates that are, in turn, eaten by predators such as sharks.

Figure VI-1: The classical marine food web as envisioned circa 1983.

by fish and sea urchins, and the energy moves quickly up the food web to the sharks. Turfs are especially common in between adjacent coral colonies. Much of the apparent grazing on corals by parrotfish that you might notice when snorkeling or diving is actually the fish eating the turf algae.[d]

In addition to the turfs, the *macroalgae*, or seaweeds,[e] on coral reefs can grow quite large if not continuously grazed. When young, seaweeds are eagerly eaten by herbivores, but when older they are tough and contain bad-tasting chemicals. Once ensconced on a reef, seaweeds are difficult to eradicate. *Together turfs and seaweeds are referred to as fleshy algae. These are the algae that take over on an algae-dominated reef.*

There is a third group of algae that are a prominent part of a healthy reef: the *crustose coralline algae*, or CCA. As their name implies, these algae form hard crusts on rocky surfaces. It is the CCA that form the sturdy, pink ridges characteristic of the surf-pounded windward reefs—important bulwarks against erosion. CCA are not as heavily grazed as other algae because they deposit calcium carbonate in the form of calcite in their cell walls. This calcite is part of the cement that holds together other reef materials.[f] *The crusts of CCA provide sites where coral larvae can settle and establish new colonies.*

Grazing by herbivorous fish and invertebrates, such as sea urchins, determines the relative abundance of the different types of algae on a reef.[g] When grazing is brisk, the fast-growing turfs are maintained as well-trimmed lawns only a millimeter or two (less than a tenth of an inch) tall; the seaweeds are kept short by frequent munching; CCA are common because the fish avoid their unappetizing, chalky flesh. However, when the number of grazers is reduced by fishing or disease, the turf

---

[d] The large Bumphead parrotfish—not apt to be overlooked as they weigh as much as a full-grown reef shark (45 kilograms, 100 pounds)—do munch on coral colonies, mostly to get at the protein-rich endolithic algae living inside the coral skeleton.

[e] The *macroalgae*, commonly referred to as seaweeds, are a diverse group of mostly multicellular marine algae that grow attached to the bottom or other solid substrates. Although we often think of algae as being small, even microscopic, macroalgae are typically measured in centimeters (inches). Kelp, a brown algae found in colder waters, can exceed 50 meters (160 feet) in length.

[f] CCA use calcite as skeletal material; corals use aragonite. Both minerals are forms of calcium carbonate. Aragonite, which makes up the struts of the reef, has a hardness of 3.5 to 4 on the Mohs scale of mineral hardness and can be scratched with a penny; the more soluble magnesium calcite, which fills in gaps and cements the reef together, has a hardness of only 2.5 to 3 and can be scratched with your fingernail.

[g] Littler, M.M., and Littler, D.S. (1984) A relative dominance model for biotic reefs. *Proceedings of the Joint Meeting of the Atlantic Reef Committee Society of Reef Studies*: 1-2.

lawns become untrimmed shrubbery and underwater forests of seaweed erupt. This process can be easily seen by caging a section of a coral reef to keep out the grazers. The turf and seaweeds flourish. This observation is an essential clue for understanding the recent decline of coral reefs. *Throughout their roughly 200 million-year reign as the dominant near-shore tropical ecosystem, coral reefs have always supported huge fish communities that have grazed down the fleshy algae.*

And now, where have all the fish gone? They've been fished out by humans. As the human population has grown, so has our demand for seafood. Moreover, we can kill more fish more quickly now thanks to technological advances. On a large scale, commercial fishing fleets supplying the global market harvest the profitable predator fish from the waters surrounding coral reefs. However, the technology need not be so grand as a factory fishing ship in order to have an impact. Simply adding a small outboard motor to a battered fishing skiff suddenly enables a local fisherman to reach a previously inaccessible reef. The larger, more desirable species are fished out first. When the sharks, groupers, and snappers are gone, fishing continues—simply moving down the food web to the herbivorous fish. Without herbivores to graze them, the fleshy algae grow larger and more abundant, overgrowing other organisms and taking over the bottom.

It is easy to picture how removing the grazers could lead to more and larger algae, but how do the algae kill corals? They have no obvious weaponry. However, algae do make the energy-rich photosynthate (sugars and carbohydrates) that is released into the water as *dissolved organic carbon* (DOC). And DOC feeds microbes. We'll come back to this shortly, but recall that each reef shark represents five short tons of algae. With one less shark, the energy that would have gone into growing and maintaining that shark is now released by the algae in the form of excess photosynthate that feeds the microbes. *So we are left with the hypothesis that killing sharks and grazers is directly fueling more microbes. More microbes mean more disease for the corals.*

⁊ § ⁊

To appreciate the role of DOC in a healthy coral reef ecosystem and the problems excess DOC can cause, we must digress for a few pages to explore the marine carbon cycle. DOC is only one of the many forms of carbon found in the oceans, but it is the important link between the algae and the microbes. Our goal here is to follow the atmospheric CO_2 as it dissolves in the sea, is captured through carbon fixation by the algae, and then either moves up the food web or is released into the water as DOC to fuel the microbes (Fig. VI-2).

The global cruise of the H.M.S. Challenger (1872-1876) marked the birth of oceanography. Challenger-based scientists routinely determined the concentration of carbon present as carbonic acid in the seawater. Ever since then, oceanographers have been diligently measuring the amount of carbon in the oceans in various forms. This is important work. There is fifty times as much carbon here as in the atmosphere, so oceanic carbon is a major factor in the global cycling of carbon.

When the CO_2 in the atmosphere meets the surface of the ocean, some of it dissolves in the seawater to form several simple carbon-containing molecules.[h] These molecules constitute the dissolved inorganic carbon (DIC) pool and it contains most of the carbon in the oceans. The different DIC molecules coexist in a dynamic equilibrium that determines not only the pH of the ocean, but also the availability of CO_2 (needed for photosynthesis) and of calcium carbonate (used by marine

---

[h] CO_2 dissolves in seawater to form carbonic acid (H_2CO_3).
(1) $CO_2 + H_2O \rightarrow H_2CO_3$
Being a weak acid, some of the carbonic acid dissociates to form bicarbonate ions (HCO_3^-) and hydrogen ions.
(2) $H_2CO_3 \rightarrow H^+ + HCO_3^-$
The hydrogen ions make the water more acidic. They also combine with the carbonate ions present in seawater to form more bicarbonate.
(3) $H^+ + CO_3^{2-} \rightarrow HCO_3^-$
At the current average ocean pH of ~8.1, approximately 90% of the DIC is in the form of bicarbonate ions, 9% as carbonate ions, and only 1% as dissolved CO_2. If the pH decreases, the amount of carbonate available will be reduced. This is a matter of concern because carbonate ions are used by corals and other organisms when depositing their calcium carbonate ($CaCO_3$) skeletons and shells.
(4) $Ca^{2+} + CO_3^{2-} \rightarrow CaCO_3$
A lower carbonate concentration makes it more difficult for the corals to deposit skeleton. Furthermore, there is a critical threshold concentration of carbonate ions below which equation 4 will be driven in the reverse direction and existing calcium carbonate skeleton will dissolve. This dissolution has a beneficial side in that it helps to buffer the increased acidity. Some of the released carbonate combines with hydrogen ions to form bicarbonate (equation 3), thus reducing the acidity. This process has helped to stabilize the pH of the oceans for millennia. The large, calcium carbonate-rich sediments underlying the oceans have provided a source of carbonate. As those deposits are consumed, further increases in atmospheric CO_2 can be expected to have a greater effect on pH.

organisms, including corals, to build skeletons and shells). The increase in CO_2 due to our burning of fossil fuels is affecting even the most remote regions of the world's oceans by shifting this equilibrium.

During photosynthesis, primary producers incorporate the carbon in the DIC pool into sugars and ultimately into new algal biomass. In oceanographic terms, the algae convert the carbon in DIC into *dissolved organic carbon* (DOC) and *particulate organic carbon* (POC). POC is defined as any organic carbon-containing particle captured by a 0.45 micron (approximately 0.000017 inch) filter. This includes cellular debris, individual cells, whales, and everything in between. Thus, the POC pool comprises all except the smallest life forms in all the oceans, along with their solid waste products and the detritus from their decay.[i] Only some microbes, viruses, and dissolved compounds pass through 0.45 micron filters. These make up the DOC.

The DOC pool is the one of greatest importance for us here, but it is also the most difficult one to characterize or measure. The usual strategy for determining the amount of DOC in a sample is to remove both the POC (the 0.45 micron filter) and the DIC (add a little acid), and then measure the remaining carbon. However, the equipment used for that measurement is notoriously unreliable. Great care also has to be taken to avoid contamination with even minute amounts of the omnipresent organic carbon compounds, such as the oils on your fingertips and vapors of boat fuel. Because it's such a pain in the butt, most marine scientists don't measure the DOC. Based on the data that we do have, the amount of carbon sequestered in *the DOC pool is immense—typically at least five times more than in the POC pool—and an amount comparable to the total atmospheric CO_2 reservoir.*

Many different sources contribute to the DOC pool, the most significant for our story being the excess photosynthate released by the al-

---

[i] In near-shore reef environments, the local POC pool can also include leaves and other detritus that wash in from the land.

gae.[j] Why do they produce more photosynthate than they need? Most often the growth of marine autotrophs, including algae, is limited by the availability of nutrients, especially phosphorous, fixed nitrogen, and iron. These nutrients are required in order to make more biomass. When an autotroph runs out of one of them, it cannot grow. However, it can—and does—continue to photosynthesize. Without an immediate need for all of the sugar for its own growth, most of the photosynthate produced is released into the environment as DOC. This may sound wasteful, but from the autotroph's point of view it is more efficient to keep photosynthesis going than to disassemble and reassemble the photosynthetic apparatus. After all, the sunlight, water, and CO_2 used in photosynthesis are all free. This release of DOC is a normal occurrence on coral reefs where nutrients are scarce. However, this gets out of hand when overfishing has reduced the fish population, including the grazers. Now the more abundant algae pump out more excess photosynthate, and this increased amount of DOC supports a larger population of microbes in the reef waters.

Until quite recently, we didn't even know that there was a significant marine microbial community. *All of oceanography from the H.M.S. Challenger expedition until the 1970s had missed more than half of the life in the world's oceans.* Marine scientists realized they were missing something important when they tried to balance the marine carbon budget. They had measured the DIC and POC pools all over the globe. They had also estimated the total amount of primary production in the world's oceans (i.e., the amount of DIC transformed into POC by photosynthesizing organisms). For the budget to balance, all of the primary production needed to be accounted for. When they compared their cal-

---

[j] DOC is also continually being formed by the breakdown of POC. Organisms that feed on POC inadvertently convert some of it to DOC. These include the super-abundant copepods, ubiquitous tiny marine crustaceans. They feed sloppily on algae, leaving "crumbs" of DOC. Microbes also feed on POC, but they have to digest it extracellularly. To do this they secrete digestive enzymes that convert the POC into seawater. However, they successfully nab only a small fraction of the DOC. The rest drifts off and contributes to the DOC pool. Even sunlight can break small POC particles apart.

Also tallied as part of the DOC are the varied exopolymers made by unicellular and multicellular organisms and released into the seawater. Exopolymers are long-chain polysaccharides that, en masse, have earned descriptive names such as mucus, slime, or goo. The surface mucus layer secreted by corals falls in this category. This DOC can spontaneously transform into POC when those long polysaccharide chains cross-link with one another through calcium bridges. Form enough of these bridges, and the DOC gels into POC. To see this for yourself, collect some seawater, filter out all the particles, then let it to sit in a bottle on the shelf for a couple days. There will be new POC.

Figure VI-2: In the early 1980s new techniques showed that most life in the oceans is microbial. These microbes, as well as their viral and protist predators, make up the Marine Microbial Food Web.

culated global primary production against the amount of known POC, the two numbers weren't even close. Almost 50% of the primary production could not be accounted for. There should have been twice as many fish in the POC pools. So where had all that carbon gone?

It turns out that the missing carbon escaped their measurement because it is in the form of DOC and DOC-fed microbes—small cells that had been missed when measuring POC. The discovery that each milliliter of seawater actually contains about half a million (5×10^5) microbial cells revolutionized oceanography. A simple calculation will show why. Collectively, there are something like 10^{28}–10^{29} microbes in the world's oceans and together they weigh as much as 2.5×10^{10} great white sharks. There are estimated to be only about 5000 great whites. This means that the microbes outweigh the most feared predator in the ocean by 10,000,000 times. In fact, microbes outweigh, outnumber, and outwork all of the other organisms in the ocean, and they make up a major fraction of marine life.

These marine microbes grow rapidly. Approximately a half million new cells are produced in every milliliter of seawater every 2 days. Since their concentration remains relatively constant, the same number of microbes must be dying. So what is killing them?

The first microbial predator guild to be identified was the protists, primarily the nanoflagellates (small single-celled eukaryotes that swim by means of flagella). These are voracious predators that zip through the ocean devouring microbes by the millions. The second predator guild was the phage, the viruses that infect and kill microbes.[k] The DOC-fueled microbes together with their protist and phage predators make up the bulk of all life in the oceans. They also form the Marine Microbial Food Web (MMFW). You can't understand coral reefs without understanding the MMFW (Fig. VI-2).

Recall that we started this journey into the marine carbon pools in order to understand how fleshy algae are able to take over on overfished reefs, replacing the corals. Our explanation is starting to take shape. ***By removing the grazers, overfishing releases the algae from top-down***

---

[k] The total number of phage on Earth is enormous, presently estimated to be a mind-boggling 10^{31}—approximately ten phage for each of their microbial prey.

control, leading to more algae. We know that algae release DOC into the reef waters—the more algae, the more DOC. Marine microbes feed on DOC—the more DOC, the more microbes. Excess DOC supports an immense community of marine microbes that could be killing corals. Thus, this model connects coral death with overfishing.

ↁ § ↁ

Many reef scientists support an alternate model to explain the co-occurrence of increased algae and coral death. They point out that both luxuriant algal growth and coral death are frequently seen on reefs that receive extra nutrients from sewage dumping, agricultural runoff, or similar sources. Perhaps the added nutrients are killing the corals directly. The need to distinguish between these two models prompted further investigation.

The first rigorous field study measuring the impact of nutrient additions was made in the 1990s on the Great Barrier Reef. This two-year study, called ENCORE (Enrichment of Nutrients on a Coral Reef Experiment), treated small patch reefs with added nitrogen and/or phosphorus. Subtle effects on the corals were observed, some positive (faster growth) and some negative (reduced reproduction). However, contrary to what some people expected, the added nutrients did not kill corals or cause the reef ecosystems to shift from a coral-dominated to an algae-dominated state.[1]

In the early 2000s, Davey Kline (Scripps Institution of Oceanography) and Neilan Kuntz (San Diego State University) set out to compare the effects of extra nutrients and DOC on coral health. For this purpose they needed a special aquarium system that could incubate hundreds of coral nubbins simultaneously, each housed in a separate chamber supplied with its own independent flow of seawater to which various test substances could be added. Such a system did not exist, so one was built and named the Aquatic Automated Dosing and Maintenance System,

---

[1] Koop, K., Booth, D., Broadbent, A., Brodie, J., Bucher, D., Capone, D. et al. (2001) ENCORE: The effect of nutrient enrichment on coral reefs. Synthesis of results and conclusions *Marine Pollution Bulletin* **42**: 91-120.

or AADAMS. Importantly, the substance being tested is added to the seawater just as it enters each of the chambers. This ensures that the microbes in the seawater do not have an opportunity to respond to or adapt to the treatment in advance. After treating thousands of coral nubbins with excess nitrogen compounds (ammonium and nitrate), phosphate, or several types of DOC (sugars and polysaccharides), the results were clear-cut. *Only the DOC treatments killed corals.*[m]

Another important clue came when Kline and Kuntz treated coral nubbins with DOC combined with a broad-spectrum antibiotic. The antibiotic protected all of the corals from the lethal effects of the DOC. Thus, microbes are also required in order for DOC to kill corals. The next step was to determine whether the DOC *produced by algae* kills corals. To investigate this, Jen Smith (Scripps Institution of Oceanography) and colleagues devised an array of two-chambered aquaria. The two chambers in each pair were separated from one another by a very fine filter (0.02 microns) that prevented the passage of microbes and viruses, but that allowed dissolved chemicals, such as DOC, to diffuse from one chamber to the other. For their experiments, they placed a coral nubbin in one chamber, seaweed in the adjoining one. In this way, the coral and seaweed were not in actual physical contact, but dissolved molecules produced by either could pass through the filter. The results? *All of the corals across a filter from seaweed died within 48 hours*. In control experiments without seaweed, all of the corals survived.

From these results we see that algae can kill corals without being in physical contact and that the algae do this by producing a substance that dissolves in the seawater. The obvious candidate is DOC that feeds the microbes already living on the corals. Consistent with this hypothesis, when a broad-spectrum antibiotic was added to the chambers, none of the corals across from algae died, showing—once again—that microbes are essential for the killing. Since the fine filter separating the coral from the algae does not allow microbes to pass through, the microbes killing the corals were not living on the algae.

---

[m] Although these experiments found that the addition of nutrients did not directly kill the coral nubbins, the results do suggest a link between added nutrients and coral death. The nutrients would stimulate algal growth. More algae would, in turn, release more DOC, and DOC does kill corals.

We know that microbes are a normal part of a healthy coral holobiont, yet the experiments just described show that these microbes can kill corals in the presence of excess DOC. What might be going on? One possible scenario is that the DOC stimulates the growth of the microbes to such an extent that the microbes consume all the local oxygen for their own respiration, thus suffocating the coral. (Recall the lethal microbial mat associated with black band disease.) To test this hypothesis, Davey Kline and Mya Breitbart (University of South Florida) measured the growth rate of the microbes. They found that the microbes on the corals dosed with excess DOC grew ten times faster than those on untreated corals.

Rapidly growing microbes consume lots of oxygen. Therefore, the next test was to measure the oxygen concentration near the surface of the corals. These measurements were made by Smith and colleagues using a microprobe that can determine differences in oxygen concentration over distances as small as one micron. Normally, oxygen levels are elevated near the surface of a coral due to oxygen production by the zooxanthellae (oxygen being a by-product of photosynthesis). When oxygen concentrations were measured on the surface of corals grown across from seaweed, there was no detectable oxygen. Further evidence: the addition of antibiotics—found previously to prevent the death of the corals—also prevents the hypoxia. ***This indicated that the rapidly growing microbes, fueled by algal DOC, are suffocating the corals.***

☙ § ❧

Algae and corals have been vying for space on the reefs for hundreds of millions of years. Until recently, the corals had always been able to rely on grazing by fish and invertebrates to keep the fleshy algae in bounds. Algae, when briskly grazed, help support the entire reef community. The munched algal biomass becomes herbivore energy and biomass, both of which move on up the food web to support the predators. When fleshy algae are consistently munched, they use most of their photosynthate to regrow their munched fronds. Take away the grazers and both the turf algae and the seaweeds grow larger. These algae then produce—and release—more DOC, thus stimulating microbial growth. When the luxuriant algae encroach on the corals, the DOC released stimulates the growth of the microbes on the coral's surface. The rapidly growing microbes use all of the local oxygen for their own respiration, and the coral polyp dies from suffocation.

Particularly insidious is the fact that DOC-induced coral death creates more living space for more algae that then release more DOC. This cycling wherein increased DOC results in the release of yet more DOC is an example of a *positive feedback loop*.[n] Unrestrained positive feedback loops essentially always cause an ecosystem to move from one stable state to another. When visualizing this using the mathematical landscape model (see pages 92-94), they propel the ball representing the ecosystem out of one valley state, over the dividing ridge, and into another stable state. *We call this positive feedback loop operating on coral reefs DDAM for DOC, Disease, Algae, and Microbes.* Driven by DDAM, a coral-dominated reef shifts rapidly to a stable, algae-dominated state.

The experiments that we have described implicate the microbes present on the coral, as opposed to some itinerant pathogens associated with the algae or some toxins produced by the algae.[o] Many microbes live in the layer of mucus on the surface of healthy corals, but their growth is kept in balance by the limited nutrients available. Might they be the killers? To explore this possibility, Kline and Kuntz collected mucus samples

---

[n] We are all familiar with *positive feedback loops*, but they are difficult to define in words. Terms that come to mind include "runaway system," "amplified effects," and "uncontrolled." As an example, consider a population of 1000 rabbits growing at a constant rate of 10% per year. The first year the population increases by 100 rabbits (10% of 1000). The following year, the same 10% rate of increase will add 110 rabbits to the population. Each year the population is larger. Thus, each year more rabbits are added to the population. If unrestrained, this generates a runaway process that is inherently unstable.

In contrast, a *negative feedback loop* damps the original perturbation. An example would be when the increased rabbit population reduces the food supply (e.g., by overgrazing), and this in turn decreases the population. Negative feedback loops are characteristic of stable systems.

[o] These experiments do not rule out other possible factors. For example, coral pathogens might also be associated with other reef organisms. Prime suspects for this role are the fleshy algae whose overgrowth on the reefs correlates with coral death. Nugues and colleagues found the pathogen responsible for a coral disease (white plague type II) present on a particular type of macroalgae. Physical contact between the algae and *Montastraea* corals was sufficient to cause the disease in the coral. Thus, a pathogen associated with the algae can take advantage of the opportunity offered when the algae encroach on the corals.

Nugues, M.M., Smith, G.W., van Hooidonk, R.J., Seabra, M.I., and Bak, R.P.M. (2004) Algal contact as a trigger for coral disease. *Ecology Letters* 7: 919-923.

Some marine algae are known to produce substances that are toxic to animals, such as the microscopic dinoflagellates that cause the notorious red tides (see page 40). It could be that the encroaching fleshy algae synthesize coral-damaging toxins, but this has not been investigated.

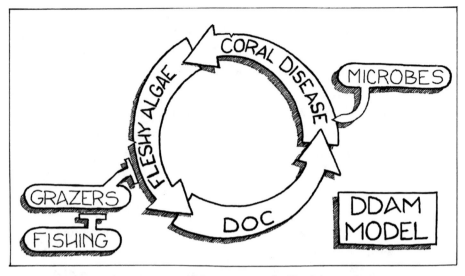

Figure VI-3: The DDAM model is a positive feedback loop wherein more algae produce more DOC that feeds more microbes. These microbes cause more coral disease and death, which in turn leads to space for more algae. Fish normally restrain this feedback loop by grazing down the algae. Overfishing removes this important governor.

from healthy colonies of four species of coral and whisked them back to the lab. They used two standard microbiological procedures to culture the microbes in the mucus. One procedure selectively supported growth of *Vibrio* spp., a group of Bacteria that includes numerous pathogens. The other method cultured enteric Bacteria, an assortment of species typically found in the human gut, including *E. coli*. Using these methods they grew large quantities of both groups of Bacteria. They then added enough of these Bacteria to the aquarium seawater around healthy coral nubbins to approximately double their normal concentration. The corals died after two weeks, showing that microbes already living on the corals could, when present in greater numbers, kill the corals.

cʒ § cʒ

The DDAM model (Fig. VI-3) explains how overfishing can lead to coral death: fewer grazing fish $\to$ more algae $\to$ more DOC $\to$ more microbes $\to$ more coral disease and death. Implicit in this model is the assertion that top-down control by grazing is sufficient to limit the growth of fleshy algae on a reef. However, in the nutrient-poor waters where coral reefs thrive, the growth of algae is often limited by the lack of necessary nutrients. When humans add more nutrients to a reef, the fleshy algae would be expected to grow faster, possibly so fast that even extensive grazing could not keep them in bounds. We explore this possibility in the next chapter when we visit two regions once renowned for their spectacular reefs.

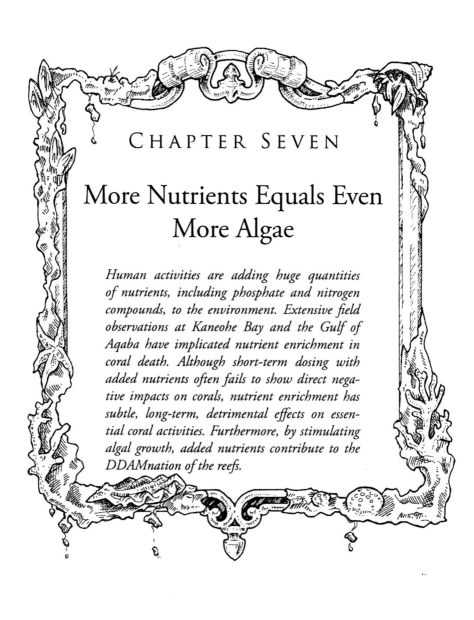

CHAPTER SEVEN

More Nutrients Equals Even More Algae

Human activities are adding huge quantities of nutrients, including phosphate and nitrogen compounds, to the environment. Extensive field observations at Kaneohe Bay and the Gulf of Aqaba have implicated nutrient enrichment in coral death. Although short-term dosing with added nutrients often fails to show direct negative impacts on corals, nutrient enrichment has subtle, long-term, detrimental effects on essential coral activities. Furthermore, by stimulating algal growth, added nutrients contribute to the DDAMnation of the reefs.

VII

The Gullible One

In order to graduate, the Worker needed to complete some work at the University of Puerto Rico (UPR) marine biology field station at La Parguera. Upon arrival, despite having taken tons of equipment and supplies with him, there were still several important items forgotten. He placed an urgent phone call to the Gullible One, "Can you bring down some reagents and equipment? And maybe help for just a little while?"

"I'll be there ASAP, but I can't stay long because of classes."

"There is one small problem—no money for a plane ticket."

"Don't worry. I can use my frequent flier miles."

"It's for her own good," thought the Worker. "She really needs to learn that bad people might take advantage of her."

Then into the phone he said, "See you tomorrow."

The Gullible One arrived cheery and was immediately delivered to the lab. She would not make it to her classes for a month. The Worker and the Gullible One set to work measuring microbial growth rates on corals collected from the reef. The results looked weird. It was the Gullible One who figured it out. "The samples can't wait until we get them back to shore. We're going to have to do the experiments on the boat."

This posed a small problem. That time of year, the trade winds are fierce in Puerto Rico. The only way to avoid getting seasick is to be underwater or on land. The worst place one could imagine was on board a boat anchored in the windy fury. So, each day after dropping anchor the Worker immediately slid into the underwater calm of the warm Caribbean waters, thereby giving the Gullible One more room to set up her mini-lab on a heaving bench. Samples of coral mucus with their associated microbes were brought to her. Precise amounts were carefully pipetted into tiny tubes while the lab bench pitched back and forth. Luckily, after the first hour the Gullible One had barfed up most of her breakfast and could better concentrate on the task at hand.

In the end, the experiments showed that the microbes on corals were growing out of control when fed extra DOC. However, they did not respond to extra nutrients such as nitrogen and phosphate. This went against all of the field observations linking nutrient enrichment to coral death. What was going on?

he Hawaii Institute of Marine Biology on the island of Oahu is one of the most beautiful marine research stations in the world. It sits, surrounded by coral reefs, on Coconut Island in Kaneohe Bay—the site of some very significant, and often unintentional, large-scale experiments in reef ecology. It was here in the 1970s that Paul Jokiel (University of Hawaii) first showed that corals would bleach when subjected to water just a few degrees warmer than normal. This early investigation was prompted by the "thermal pollution" from a newly-constructed power generating station that dumped hot water into the bay. At that time, thermal pollution was considered an unusual and highly localized stress affecting only a few coral reefs. It was not until the 1980s that widespread coral bleaching was documented, and it was still later that the potential impact of global climate change on coral reefs worldwide was fully realized.

Kaneohe Bay also provided a classic example of the effects of a local stressor: sewage discharge. After World War II, Kaneohe became a burgeoning bedroom community for the rapidly growing town of Honolulu. Its population swelled from about 5,000 to almost 30,000 by 1960, and doubled again by 1990. This many people produced a lot of sewage. Treatment plants were built that then discharged the treated water into the bay. Similarly, treated sewage from the adjacent Kaneohe Marine Corps Air Station had been dumped into the bay since the 1940s. By the 1970s, the corals of Kaneohe Bay were dying and luxuriant "green bubble algae" were visibly overgrowing the reefs. Some areas of the bay, once described as "coral gardens," were now pronounced "dead." Then, between 1977 and 1978, most of the sewage was diverted to a deep ocean outfall offshore. Stephen V. Smith (University of Hawaii) and colleagues seized the opportunity offered by this Kaneohe Sewage Diversion Experiment and documented the subsequent shift of the bay from an algae-dominated ecosystem back to one dominated by corals. This heartening story is a clear example of how reefs can recover when relieved of a noxious local stressor.

What was in the "treated" sewage that killed the corals in Kaneohe Bay? The most obvious candidates are the nutrients, including nitrogen (nitrates or ammonium), phosphate, and iron. Since nutrients are often growth-limiting on coral reefs, their addition from human activities can have a dramatic impact on the whole ecosystem. The main trouble from nutrient enrichment is stimulated algal growth, but, as we'll see, nutrients can also directly affect other organisms. Industrial production has dramatically increased the amount of usable nitrogen in the world's oceans. This warrants a brief digression, as the added nitrogen has had a significant impact on coral reefs.

About eighty percent of the air that we breathe is gaseous nitrogen (N_2). The two nitrogen atoms are so strongly joined together by triple bonds that they cannot be used as a nitrogen source by most organisms. Only a few specialized Bacteria are able to "fix nitrogen," i.e., to break these bonds and combine the nitrogen with other elements to form ammonium and the other nitrogen-containing compounds necessary for life.[a]

Natural sources of fixed nitrogen have been highly valued since the beginning of agriculture. In the Middle Ages, field rotation systems were introduced that included a legume crop (e.g., clover, alfalfa, beans). Legumes have nitrogen-fixing Bacteria living symbiotically in their roots that enrich the soil. During the early industrial era, population surges were spawned by—and then required—additional nitrogen sources. From 1840 to 1880 the source was guano, the nitrate- and phosphate-rich material produced over centuries by the accumulation in dry climates of shit (literally) from millions of seabirds (and sometimes bats). In the ensuing race to exploit this limited resource, guano mining became a highly competitive industry. To give Americans an edge, in 1856 the United States Congress passed the Guano Islands Act that allowed USA citizens to take possession of unclaimed islands and mine the guano deposits for use by USA interests only. This led to claims and subsequent guano mining on islands that had long served as seabird nesting and roosting havens, including some in the Line Islands.

All of this changed in the early 1900s when a German chemist, Fritz Haber, devised a chemical method that "fixed" nitrogen in the labora-

---

[a] We use *nitrogen fixation* to refer to only the biological process, although as sometimes defined it also includes abiotic means such as lightning and industrial processes. Lightning converts N_2 to nitrates, accounting for about 5% of the total nitrogen fixed by all means each year.

tory. By 1913, this process had been scaled up to industrial-level production by Carl Bosch. During World War I, the industrialized Haber-Bosch process provided the nitrogen compounds the Germans needed to manufacture ammunition without dependence on guano supplies overseas—a capability essential for their sustained war effort.

The Haber-Bosch process continues to play a central role in today's world, now as the primary source of nitrogen for the fertilizers that underpin modern agriculture. The process is extremely energy intensive and also uses natural gas as a feed-stock. Stimulated by the increased need for fertilizer and by cheap fossil fuels, the amount of fixed nitrogen added to the biosphere by humans greatly exceeds natural terrestrial production and approaches the total for terrestrial and ocean systems combined. Humanity now controls the Earth's nitrogen cycle.[b]

Like fixed nitrogen, phosphate is one of the primary plant nutrients and a major component of chemical fertilizers. Prior to the mid 1800s, farmers were dependent on natural phosphate sources such as guano. That changed in 1842 when an Englishman, John Bennett Lawes, obtained a patent on his process that uses sulfuric acid to convert insoluble phosphate-containing ores into "superphosphate," a superior plant fertilizer. Ever since, we have been voraciously mining high-phosphate ores to manufacture fertilizers.

Our modern agriculture depends on continual massive inputs of chemical fertilizers containing industrially-produced nitrogen and mined phosphate, combined with iron and other trace minerals. Nitrogen is abundant in the atmosphere, so the potential limitation here is having sufficient cheap energy to convert it to an affordable usable form. In contrast, mineable phosphate ores are a limited resource that accumu-

---

[b] The calculated values for the amounts of nitrogen added to the biosphere by human activities and by natural processes vary, often due to differences in inclusivity, but the overall conclusion is always the same. Humans are grossly altering the global nitrogen cycle. In the 1990s, the nitrogen used in agriculture and added to the atmosphere by the burning of fossil fuels exceeded 160 teragrams (Tg) N per year. In comparison, 110 Tg are produced by terrestrial biological nitrogen fixation and 140 Tg by nitrogen fixation in marine environments.
  Gruber, N., and Galloway, J.N. (2008) An Earth-system perspective of the global nitrogen cycle. *Nature* **451**: 293-296.
  A more comprehensive measure of the anthropogenic nitrogen added to the biosphere also includes the burning of biomass, replacement of natural ecosystems with legume crops, and other land use changes. On this basis, anthropogenic sources totaled 210 Tg per year in the 1990s.
  Vitousek, P.M., Aber, J.D., Howarth, R.W., Likens, G.E., Matson, P.A., Schindler, D.W. et al. (1997) Human alteration of the global nitrogen cycle: sources and consequences. *Ecological Applications* **7**: 737-750.

lated over vast periods of time and are found only in particular regions of the world—USA (mainly Florida), China, Israel, Jordan, and Morocco. There is no substitute for phosphate in food production. While the looming specter of "peak oil" has garnered a lot of attention, world peak phosphate production approaches or may have already quietly passed, depending on which authority you consult. All agree, however, that our food supply is dependent on this vanishing resource. Furthermore, phosphate may well be the Achilles heel limiting biofuel production. Whatever plant or algal source you envision as the feedstock for bioconversion into fuel, growth of that source will require phosphate, and lots of it.

In this light, it makes no sense to be wasting fertilizers. As if that weren't enough to condemn its over-application, this habit combined with poor soil conservation practices has led to the massive influx of fixed nitrogen, phosphate, and other nutrients into most aquatic environments. This careless and wasteful environmental "fertilization" has transformed our lakes, rivers, and near-shore oceans into algal-microbial slimes. One particularly visible example is the dead zone in the Gulf of Mexico, an area the size of the state of New Jersey. Here the runoff from fertilizer applied by agribusiness throughout the American Midwest spews into the gulf at the mouth of the Mississippi River. Dead zones are now a global phenomenon, with a 2008 report tallying 405 worldwide. Some of the agricultural nitrogen also finds its way into our diet, only to be later released into rivers and oceans as sewage. Whatever the route, this is a clear instance of too much of a good thing bringing environmental devastation.

⁋ § ⁋

For one of the premier SCUBA diving experiences of your life, head to the Gulf of Aqaba (aka the Gulf of Eilat), a portion of the Red Sea bordered by Jordan, Israel, Egypt, and Saudi Arabia. Some of the world's most stunning reefs are to be found here. Visibility in the clear water is more than 60 meters (200 feet); spectacular reefs are often located right off the shore. A short ride by camel brings you and your dive gear to the coast. When you get to a good spot, just walk into the water and immerse yourself in all the wonderful life. But don't postpone this trip too long. This natural wonder is being killed by nutrient enrichment from aquaculture.

Originally aquaculture was perceived as a benign and innovative way to feed people from the sea without depleting wild fisheries. Production by these farms has increased to the point where, as of 2009, it accounts for half of the total fish and shellfish in our diet. However, aquaculture as currently practiced has some major flaws. Farmed fish are often fed fishmeal and fish oil, both products from wild fisheries. Excess feed that drops through the cages, combined with the fish poop, adds nutrients to local waters. This is now happening in the Gulf of Aqaba.

At the northern tip of the gulf sit the Israeli port of Eilat and the Jordanian port of Aqaba (famously seized by the forces of the Arab Revolt led by Lawrence of Arabia during World War I). By 1990, coral cover on the reefs in this region had declined markedly, due to the discharge of untreated sewage from both cities. Aqaba then began treating their sewage and diverting the nutrient-rich outflow to inland landscape irrigation and Eilat followed suit in 1995. As a result, nutrient enrichment from sewage was greatly reduced, only to be replaced in the early 2000s by nutrients from fish pens installed for raising gilthead sea bream *Sparus aurata*, a non-indigenous species. By 2000, this aquaculture operation was estimated to release dissolved organic matter containing 240 tons of nitrates and 40 tons of phosphates each year—approximately 97% of the total nitrogen and 66% of the total phosphate added to the northern Gulf of Aqaba.[c] Because fish production has increased by 50% since then, these figures are now underestimates. As a result, nutrient levels in the gulf remain elevated, the incidence of coral disease is on the rise, and the deterioration of the unique reefs of the gulf goes on.[d]

Both Kaneohe Bay and the Gulf of Aqaba clearly link excess nutrients and coral decline. There are other examples demonstrating similar

---

[c] The rest of the phosphate arrives as phosphate dust, a byproduct of the loading of locally-mined phosphate onto ships in the gulf ports.
Loya, Y. (2007) How to influence environmental decision makers? The case of Eilat (Red Sea) coral reefs. *Journal of Experimental Marine Biology and Ecology* **344**: 35-53.

[d] Loya, Y., Lubinevsky, H., Rosenfeld, M., and Kramarsky-Winter, E. (2004) Nutrient enrichment caused by in situ fish farms at Eilat, Red Sea is detrimental to coral reproduction. *Marine Pollution Bulletin* **49**: 344-353.

correlations from around the world. However, the experimental addition of excess inorganic nutrients to corals in aquaria and in the field typically has no apparent detrimental effects on corals. These experiments include those by Kline and Kuntz who treated coral nubbins with excess nitrogen compounds (ammonium and nitrate) or phosphate using the AADAMS flow-through system (aka The Bitch, see pages 110-111). The inorganic nutrient treatments did not kill the corals, whereas treatment with DOC did. Likewise, ENCORE, the two-year field study in which nutrients were experimentally added to corals on the Great Barrier Reef, found no clear link between nutrient enrichment and coral death. Other experiments have yielded similar observations. Why the discrepancy between these detailed studies and the observed decline of nutrient-burdened coral reef ecosystems?

Even if increased nutrients do not kill corals directly, they do promote growth of the algae. The resulting increased algal biomass produces more excess photosynthate that is released into the water in the form of DOC. This additional DOC supports more microbes and enables them to grow faster. The rapidly growing microbes on the surface of the corals consume all of the local oxygen and suffocate the coral animal. This leads to more coral disease, hence the observed correlation between nutrient-driven algal blooms and coral death. This is the DDAM model again (Fig. VII-1).

The increased algal growth due to nutrient enrichment can be countered to some extent by top-down control by grazing. Thus, added nutrients would be expected to have less impact on corals if the reef has an abundant and diverse population of fish, including herbivorous grazers. Unfortunately, since both nutrient enrichment and overfishing occur on reefs near humans, they often act synergistically to kill the reef.

ᔐ § ᔐ

Even though many experiments have shown that nutrient enrichment alone does not kill corals outright, they do not preclude more subtle effects undermining the long-term survival of corals. For example, field studies spanning two years or more in the Gulf of Aqaba and on the GBR

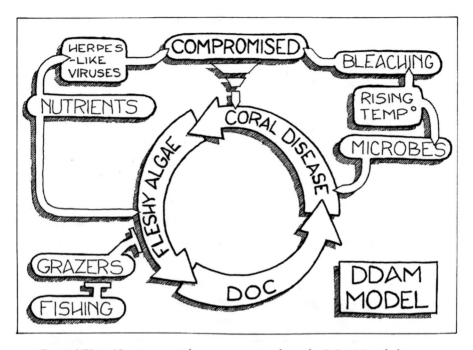

Figure VII-1: Nutrient enrichment can exacerbate the DDAM cycle by fertilizing the reef, thereby removing bottom-up control of algal growth. Added nutrients may also compromise the holobiont's immune defenses as evidenced by the reactivation of latent herpes-like viruses. Increased temperature makes all of this worse for the coral by disrupting their delicate relationship with their zooxanthellae and by increasing microbial growth rates.

found that nutrient enrichment diminishes coral reproduction.[e] Earlier field studies on the Great Barrier Reef also linked phosphate enrichment to decreased reef calcification.[f] These effects vary depending on species, nutrient concentrations, and other factors, but suggest a link between nutrient addition and lower rates of calcification by the corals.

---

[e] Loya, Y., Lubinevsky, H., Rosenfeld, M., and Kramarsky-Winter, E. (2004) Nutrient enrichment caused by in situ fish farms at Eilat, Red Sea is detrimental to coral reproduction. *Marine Pollution Bulletin* **49**: 344-353.

Koop, K., Booth, D., Broadbent, A., Brodie, J., Bucher, D., Capone, D. et al. (2001) ENCORE: The effect of nutrient enrichment on coral reefs. Synthesis of results and conclusions *Marine Pollution Bulletin* **42**: 91-120.

[f] Kinsey, D.W., and Davies, P.J. (1979) Effects of elevated nitrogen and phosphorus on coral reef growth. *Limnology and Oceanography* **24**: 935-940

Nutrient additions can also disrupt the coral-zooxanthellae symbiosis. Normally, the coral limits the growth of their zooxanthellae by restricting their supply of fixed nitrogen.[g] Thus enslaved, the zooxanthellae keep less of the photosynthate for their own use and transfer more to the coral, fueling robust reef building by the corals. Nutrient enrichment frees them from host control, allowing them to grow more rapidly. Under these conditions, their total photosynthesis is unchanged, but they use more of the photosynthate themselves, leaving less to support reef construction by the coral.

Nutrients also affect the other symbioses within the holobiont. Recall that Linda Wegley Kelly used metagenomics (see pages 66-67) to show that at least five different types of organisms work together to produce and recycle nitrogen compounds. Extra nitrogen would be expected to disrupt these finely-tuned relationships. This possibility is being actively investigated by coral microbiologists.

Nutrient enrichment also aids and abets coral pathogens. Outbreaks of coral disease have been reported near sewage outflows. Few studies have tested the effect of nutrients on the incidence or severity of specific diseases. However, a study of black band disease by Joshua Voss and Laurie Richardson (both at Florida International University) found that nutrients accelerated the disease's advance both in the field and in aquaria.

Metagenomics has also been used to monitor the effects of nutrient enrichment on the holobiont. This is important because some coral-associated microbes might be potential pathogens while others help to stave off disease.[h] In her aquaria experiments, Becky Thurber found that added nutrients shifted the community to one dominated by microbes typically found on diseased corals (see pages 82-84). The coral-associated viruses also dramatically changed, most notably with a pronounced increase in the number of herpes-like viruses. This finding suggests that the nutrients cause reactivation of latent herpes infections, possibly by suppressing the coral immune system.

---

[g] For details, see footnote r on page 68.
Falkowski, P.G., Dubinsky, Z., Muscatine, L., and McCloskey, L. (1993) Population control in symbiotic corals. *BioScience* **43**: 606-611.

[h] Rypien, K.L., Ward, J.R., and Azam, F. (2009) Antagonistic interactions among coral-associated bacteria. *Environmental Microbiology* **12**: 28-39.

ოა § ოა

The diverse effects of nutrient enrichment—increased algae, disruption of the holobiont, and reactivation of latent viruses—make this local impact a major threat to coral reefs. When the same coral reefs are subjected to overfishing, as is often the case, the enhanced algal growth cannot be effectively countered by grazing fish and the potential for devastation is even greater. We know that healthy coral reefs can exist under different nutrient conditions, i.e., in different regions of biogeochemical space (see pages 91-92). Nutrient concentrations on some reefs might naturally be so low that additions can be tolerated without triggering an algal bloom, even with a reduced fish population. However, most reefs appear to need the fish or invertebrate grazers (such as the sea urchins) to keep the fleshy algae under control. Adding extra nutrients gives the algae an advantage. Add enough nutrients, and even abundant grazers cannot save the coral. With apologies to all of the wonderful phycologists of the world, it is the DDAMned algae!

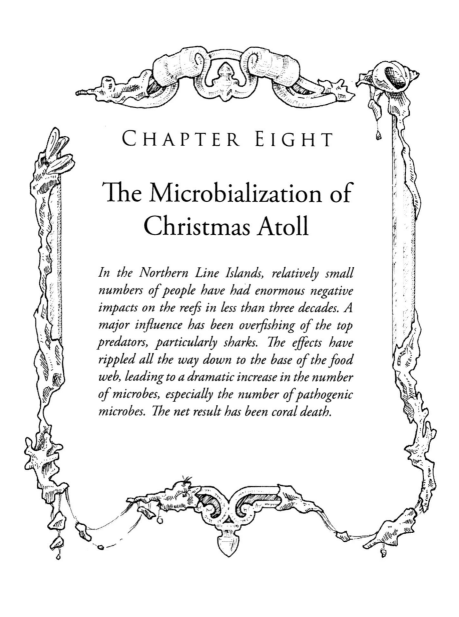

CHAPTER EIGHT

The Microbialization of Christmas Atoll

In the Northern Line Islands, relatively small numbers of people have had enormous negative impacts on the reefs in less than three decades. A major influence has been overfishing of the top predators, particularly sharks. The effects have rippled all the way down to the base of the food web, leading to a dramatic increase in the number of microbes, especially the number of pathogenic microbes. The net result has been coral death.

VIII

Sampling Sewage

One day it occurred to the Worker that instead of using purified chemicals to simulate the nutrient addition caused by sewage dumping, it might make sense to use, well, real sewage. Bocas del Toro, Panama, does have a sewage treatment plant, and much of the sewage from the town does, in fact, pass through it. However the local magistrates find it far more profitable to sell the chemicals intended for use in the treatment process and let the untreated sewage disappear into the waters of the adjacent reefs. These forward-looking individuals thus provided the Worker with the needed raw sewage, and for that we thank them.

A majority vote elected the Misdirected as the collector of raw sewage. Originally, the majority thought that snorkeling to the middle of the pond would be the best method of collection, as well as hilarious entertainment. The Misdirected's whining when presented with this plan, however, was too much to bear. So on to Plan B. In the tradition of field biologists worldwide, a prototype sampler consisting of a plastic beaker duct-taped to a PVC pole was constructed. This sampler did indeed work, however the floaters, condoms, etc. in the sample were not compatible with The Bitch's Venturi valves.

Obviously, a more refined sampler was called for. In Version 2.0, the undiscriminating beaker was upgraded to a bleach bottle with a ten-centimeter (four-inch) round hole cut near the top. Nitex, a nylon mesh with 100-micron pores, was glued over the hole. When lowered into the pond, sewage, sans larger debris, percolated into the bottle. When coral nubbins in The Bitch were dosed with this sewage soup, the corals started to die.

Even though it may seem obvious that sewage kills things, it has been difficult to conclusively link various human-derived stressors, such as sewage, to coral death. And small doubts can be used by naysayers to avoid doing anything. This is the sort of rationalization that lets the powers-to-be in Bocas del Toro pour untreated sewage onto their reefs, even though they eat the fish from those waters and even though tourism is one of the town's main sources of income. That's why, in addition to characterizing a pristine coral reef, a major goal of the Northern Line Islands Expedition was to document what happens when a relatively small number of people use a coral reef in an unsustainable manner.

ntil recently, the most complete case studies connecting local impacts to reef collapse had been from Jamaica's Discovery Bay and Hawaii's Kaneohe Bay. However, both stories are incomplete. For both areas, the impact from fishing preceded the studies by centuries, beginning in the pre-colonial days in Jamaica and upon the arrival of the Polynesians in Hawaii. Since we lack detailed data on the reefs prior to human intervention, interpretation must rely on our extrapolation of fish populations backwards in time based on anecdotal reports and best estimates. Even the more recent scientific data that we do have is seriously incomplete as it says nothing about what the microbial communities were doing during the ecosystem shifts. Although researchers were monitoring the fish, the sea urchins, and the algae, no one was paying any attention to the microbes until the 1980s. Only recently have we come to realize that it's impossible to understand what is happening to coral reefs—or any other ecosystem—without considering the microbes and their role in the food web. The metagenomic tools required for investigating the microbial and viral communities weren't even invented until the 21st century.

Now the tools needed to observe all the players, including the microbes, are in place, but we can't go back in time and re-monitor the conditions at Discovery or Kaneohe Bay. However, we can survey similar reefs in varying stages of decline today, essentially substituting space for time. For this purpose, we need several reefs located in the same region that have been impacted by humans only recently, and to differing degrees. Just such a gradient of reefs, ranging from nearly pristine to greatly altered, is provided by four of the Northern Line Islands: Kingman, Palmyra, Fanning, and Christmas.

Enric Sala and Stuart Sandin, two of the Fish from the Scripps Institution of Oceanography, recognized the unique opportunity presented by these four atolls. They also realized that seizing this opportunity required assembling an expedition that would include both classical reef biologists and microbial ecologists: thus, the Benthics, the Fish, and the Microbes on board the first Northern Line Islands Expedition.

Sala, a marine ecologist, is a modern day explorer who bears a striking resemblance to the movie star Antonio Banderas. He did much of the

Enric Sala

fund raising and planning for the expedition. Once the expedition was underway, he served as the Chief Scientific Officer on board, charged with responsibility for keeping everyone rowing in the same direction—not a simple task in the close quarters of the crowded White Holly.

In the great tradition of Academia, it was Sandin, a post-doc working with Sala, whose job it was to actually get everything done. This included contracting and retrofitting the White Holly, purchasing clipboards and pencils for fish counting, and thousands of other details.

The expedition followed in the wake of earlier explorers. Archaeological evidence shows that the Northern Line Islands were first visited by voyaging Polynesians who used the atolls for several thousand years as sites for brief stopovers or short-term settlements. All four atolls were uninhabited when rediscovered by Cook and Fanning in the late 1700s. Once they had been added to the nautical charts, they were frequented by whalers and other vessels, the crews harvesting the abundant birds, fish, and sea turtles.

Christmas Atoll was claimed by the USA in 1856 for guano mining, but very little mining ever took place. The first permanent settlers arrived in 1882 to fish and grow coconut palms for copra production.[a] During World War II, the occupying Allied forces constructed the first airstrip. With the Cold War came extensive atmospheric nuclear bomb testing by both Great Britain and the USA (1956 to 1962). Accompanying that program was a construction force of approximately 4000 workers to build the needed infrastructure, including roads and runways. After-

wards, when testing indicated no radioactive contamination of the land, copra production was resumed.

The smaller Fanning Atoll was formally annexed by Great Britain in 1888. The new owners proceeded to blast a deep passage into the lagoon on the west side of the atoll and then promptly dubbed it the English Channel. From 1902 until 1964, Fanning hosted a cable station on the then-vital Trans-Pacific telephone cable.

Stuart Sandin

In 1979, Christmas and Fanning were united with the Southern Line Islands, the Gilbert Islands, and the Phoenix Islands to form the new Republic of Kiribati. Most of the land of Kiribati is less than two meters (six feet) above sea level; most of the 110,000 people live in the crowded Gilbert Islands. Christmas constitutes more than 70% of the total land area of the republic—while measuring a mere 322 square kilometers (124 square miles). Beginning in 1988, the government promoted "voluntary" resettlement of 10% of the population from the Gilbert Islands to Christmas or Fanning. As a result of this policy, Fanning and Christmas Atolls, both of which had populations of about 450 in 1987, now are home to 2,500 and over 5,000, respectively. Further population increases are expected.

---

[a] Copra is the dried meat of the coconut, which is then crushed to extract the coconut oil.

Kingman Reef and Palmyra Atoll were also discovered in 1798 by Captain Fanning. Since the amount of dry land on Kingman is miniscule (less than 0.01 square kilometers or 2.5 acres), human impact has been limited to brief usage of the lagoon as a refueling station for commercial flying boats plying a trans-Pacific route in 1937-38. The reef is now an unincorporated territory of the USA and is closed to the public.

Palmyra, with 2.75 square kilometers (680 acres) above water, has a mixed history of public and private ownership. The US Navy took over the atoll during World War II for use as an airbase. They dredged a channel so that ships could enter the protected lagoon, then provided more shelter by building dikes across the lagoon to reduce water flow. Before leaving, the military dumped several tons of ammunition and ordnance into one of the ponds created by the dikes. Coral rubble was also bulldozed to construct a long airstrip. During the post-war years, Palmyra was once again in the hands of private owners who rejected offers from commercial interests seeking to develop the atoll, as well as a proposal by the USA government to make of it a nuclear waste dump. The atoll was purchased by the Nature Conservancy in 2000 with the intent of insuring its preservation. In 2001, the atoll along with 2,085 square kilometers (515,232 acres) of surrounding waters and coral reef were designated the Palmyra Atoll National Wildlife Refuge under the U.S. Fish and Wildlife Service. In 2009, Palmyra was incorporated into a new marine national monument along with Kingman.[b]

℘ § ℘

For the researchers on the Northern Line Islands Expedition, Kingman Reef to the north represented a near-pristine coral reef. Moving south-

---

[b] A fifth atoll, Teraina or Washington Atoll, lies between Fanning and Palmyra. Geologically, Washington is similar to the other Northern Line Islands. At one time it boasted the typical coral reefs surrounding a central, moderately hypersaline lagoon. However, a few centuries ago a major global precipitation band shifted. Washington now has far more rain and less evaporation than the other Line Islands (4,000 mm or 160 inches of rain per year, more than four times as much as falls on Christmas). Today it is a densely wooded island less than half the size of Fanning but inhabited by six hundred people and supporting an active copra industry. In place of the typical lagoon, one finds tilapia fish swimming in a large freshwater lake, the only freshwater lake in the Line Islands. On the 2005 expedition to the Northern Line Islands there was not enough time to thoroughly investigate this fifth atoll. Our few snorkels and dives there revealed that most of the larger fish have been fished out, much of the coral is dead, and algal patches are abundant. We will be revisiting Washington in 2010 to complete a proper survey of this interesting island.

Figure VIII-1: The gradient of human population and its effects on reef health in the Northern Line Islands. Where there are more people, there are more pathogen-like microbes, more fleshy algae, and more coral disease. Conversely, fewer people mean more sharks and more healthy coral

ward to Palmyra, Fanning, and then Christmas Atoll, one travels along a gradient of increasing local human impact (Fig. VIII-1). Photographs and measurements by the Benthics documented the accompanying shift on the bottom from coral- to algae-domination. Live coral covers 40-60% of the substrate on Kingman, but only about 15% on Christmas. CCA also decline dramatically from Kingman to Christmas, whereas both turf algae and seaweeds increase in abundance and size.

The loss of coral from Christmas and Fanning is recent, reflecting the increased colonization of these atolls after 1988. Benthic surveys of Christmas made just 15 years ago recorded abundant corals. Today, on the decimated reefs stand coral skeletons, upright and silent, not yet felled by storms or the incessant waves. All the evidence confirms that the coral die-off on Christmas (and on Fanning, as well) is the result of less than two decades of increased local human impacts.

How can a few thousand people cause such widespread devastation so quickly? One possibility might be by discharging nutrient-laden sewage into the reef waters. However, there are simply not enough humans on these atolls to cause measurable changes in the nutrient levels.[c] *In contrast, there are more than enough determined fishermen to dramatically reduce the communities of grazers and to kill all of the sharks.* Sharks are particularly easy to catch; you don't even have to hunt them down. One person throwing out chum from a boat will attract every shark within a large radius. Or you can gillnet them at the entrances to the lagoon. This has been the story on Christmas and Fanning.

Sharks are not the best tasting fish in the ocean; their meat is not worth much in the marketplace. Unfortunately, their fins are highly valued for shark fin soup, a cure for flaccid male egos.[d] Shark fins bring between US$100 and US$5,000 per kilogram (2.2 pounds) in Singapore, Hong Kong, and Taiwan. Imagine that you've been "voluntarily" resettled on a remote atoll in the Pacific. There are no jobs, you have no

---

[c] The Microbes measured nutrient levels in the waters surrounding all four atolls. The observed nutrient concentrations follow the patterns expected for this region. Most telling, measurements of the stable isotopes of nitrogen that can very precisely pin-point human-derived sewage provided no evidence for sewage being a major factor in the coral decline. This does not mean that sewage is not a problem in other areas of the world, rather that it is not a justifiable explanation for the coral death occurring in the Northern Line Islands.

[d] Ironically, due to the high mercury content found in shark fins, consumption of shark fin soup can cause male sterility.

money, but you have been given a boat for fishing. Wouldn't you go shark finning?

Sharks were once abundant on both Christmas and Fanning. Early explorers left vivid accounts of their large numbers (see page 19), and those early observations have been confirmed by more recent surveys. Today, there are essentially no sharks to be found on either of those atolls. In contrast, on Kingman and Palmyra most of the fish biomass is swimming about as sharks and other apex predators such as vampire snappers. For simplicity, we use "sharks" here as a proxy for the apex predators as a whole.

The results of overfishing the grazers are easy to picture. Without the herbivores, the fleshy algae will grow larger and release more DOC, thus fueling the DDAM cycle. But more fish than just the grazers are required for a healthy reef. Top predators are also essential. The reason why is not intuitive and has confused many, including coral reef scientists. However, it is vitally important, so we will go through it step-by-step. Recall that any ecosystem can be envisioned as an energy pyramid made up of trophic levels, with the primary producers at the bottom (see pages 100-101). Above the primary producers, the organisms in each level feed on the level just below them. The greatest amount of energy is always in the primary producers at the base; approximately 10% of the energy at each level can be available to be drawn up by the organisms in the next higher level.

One can instead build a pyramid representing the biomass in each trophic level. The trophic levels remain the same: the primary producers make up the first level, the herbivores the second level, and then on up through the various levels of predators to the apex predators at the top. You might expect that since the base of the pyramid contains the most energy, it would also contain the greatest biomass. That is indeed what had been observed for most ecosystems, including supposedly "pristine" coral reefs. However, those coral reefs had already been degraded by fishing and perhaps other human impacts. Due to shifted baselines, no one realized that those bottom-heavy fish biomass pyramids did not represent pristine coral reefs.

The first data exposing this error was collected by Alan Friedlander

(Hawaii Cooperative Fishery Research Unit; University of Hawaii) and Edward DeMartini (National Oceanic and Atmospheric Administration Fisheries Service), two of the Fish. In their earlier surveys of the Northwestern Hawaiian Islands, they had discovered that there was more biomass in the apex predators (i.e., the sharks, giant trevally, and groupers) than in all of the other fish lower on the pyramid combined. The fish biomass pyramid for pristine coral reefs is inverted.

How can there be more biomass at the top if there is more energy at the bottom? To answer this, we have to take a closer look at the interplay between algae, grazers, and predators on a pristine reef. Here intense grazing keeps the turfs and fleshy algae closely cropped and in the process rapidly draws energy up to the second trophic level. When there is less algal biomass, less photosynthate is produced and more of that photosynthate is used by the algae themselves to regrow the munched fronds needed to keep up with the incessant grazing. Hence very little photosynthate is released as DOC to feed the microbes.

The herbivorous fish grazing the algae are themselves under intense predation, so a lot of algal energy goes into replacing the eaten grazers. It is much more energetically expensive to raise new fish than it is to maintain adult fish. (Anyone with kids knows this principle!) In this way, the energy moves up the pyramid, level after level, all the way to the apex predators at the top. When there are no humans fishing the reef, the community of apex predators is huge, all hungrily drawing up energy from below, thus leaving less energy available to feed the microbes.

The total amount of fish biomass on a pristine coral reef is much greater than on a fished one, and most of that greater biomass is in the apex predators. Removing many of them dramatically reduces the total biomass. When the large predators have been fished out, there are more grazers, but not as many as expected. Although the grazers are not being eaten, their numbers are limited by other factors.

The large predators also control the amount of reef algae by another, less obvious means. Some species of herbivorous fish can grow quite large—large enough to escape most predation—whereas smaller species are always on the menu. When there are lots of predators, the large herbivores have an advantage over the small ones. Take away the predators, and the smaller herbivores dominate. This correlation between the size of the herbivorous fish and the number of large predators was confirmed by observations in the Northern Line Islands. This is important because the big herbivorous fish are better at controlling the fleshy algae; they literally take bigger bites. Nibbling by the smaller fish is not as effective, especially with the larger seaweeds. Thus, fishing out the large predators favors small—rather than large—herbivorous fish and these smaller grazers can't control the fleshy algae.

When the fleshy algae are not effectively grazed, there is, of course, more algal biomass producing photosynthate. The situation is worsened because algae that are not grazed need less photosynthate since they do not need to replace munched fronds. Both of these factors contribute to the release of more photosynthate into the water as DOC. This is a boon for the microbes. Whereas a pristine coral reef is characterized by abundant sharks, well-trimmed algae, lots of coral, and few microbes, an overfished reef is dominated by an algal-microbial slime. These two contrasting reef states are exactly what the expedition found in the Northern Line Islands. *The Fish found that 85% of the total fish biomass on Kingman was apex predators—the highest percentage recorded anywhere to date.*[e] *At Christmas (the atoll with the largest human population of the four visited), apex predator biomass fell to 19%.*

The DDAM model predicts that overfishing—especially the removal of the apex predators—will allow the microbes to become abundant and grow fast, fueled by the DOC from the ungrazed fleshy algae. *Again, this is exactly what was observed in the Northern Line Islands. There are ten times as many microbes in the reef waters at Christmas as at Kingman. There are also ten times as many virus-like particles* [f] *on Christmas.* (These virus-like particles are most likely phage preying on the larger microbial community.)

Even more disturbing than the increase in microbial numbers was the shift in the types of microbes between Kingman and Christmas. The ex-

---

[e] We later measured even greater predator biomass on Starbuck Island in the Southern Line Islands, but this data is still being analyzed.

[f] *Virus-like particles* are particles tentatively identified as viruses based on their size, shape, and other physical properties. However, they have not yet been further characterized or shown to be able to infect a cell and replicate therein.

pedition collected samples at each atoll for metagenomic analysis of both the viral and microbial communities. For this operation, the Microbes employed their Supreme Sucker to collect 100-120 liters (approximately 30 gallons) of seawater from close to the reef surface. The microbes and viruses in the seawater were collected on a filter with very small (100 kDa) pores. The microbes were then separated from the viruses by catching the microbes on a 0.45 micron filter. The viruses passed through that filter and were further purified by a centrifugation step to separate them from any small microbes or free DNA. The DNA was extracted from the collected microbes and viruses, sequenced, and analyzed as described earlier (see pages 63-66).

The results were striking. The smaller microbial community at Kingman ekes out a modest living, about half of them being self-sufficient autotrophs; the multitudes at Christmas are mainly heterotrophs that depend on other organisms to produce their food—primarily the DOC donated by the ungrazed fleshy algae.

Furthermore, the heterotrophic microbes on Christmas are not your commonplace marine heterotrophs. Rather they are "super-heterotrophs" that prefer to live in very plush, energy-rich environments. Since the most energy-rich environments are within the bodies of animals and plants, super-heterotrophs are essentially pathogens. Many are closely related to well-known pathogens such as *E. coli*, *Streptococcus* spp., *Vibrio* spp.[g] , *Staphylococcus* spp., and numerous plant pathogens. *The seawater on Christmas is swarming with pathogens. You would not want to swim here. Unfortunately, the corals have no choice.*

The changes observed at Christmas relative to Kingman—a ten-fold increase in the number of microbes and viruses, as well as a shift to more "pathogenic" microbes—echo the shift observed toward more pathogenic microbes when corals were subjected to various stressors in Thurber's aquarium studies (see pages 82-84).

---

[g] *Vibrio* is a genus of Bacteria that includes many pathogenic species. There is a very quick and simple method for assaying the number of vibrios in a sample such as an aliquot of seawater: prepare sterile agar culture plates made with a growth medium containing bile salts, use your sample to inoculate the plates, and then count the resultant bacterial colonies. Since vibrios can tolerate bile salts but most other microbes cannot, this gives you an estimate of the number of vibrios present. Based on this assay, we found that there are at least a thousand times more vibrios at Christmas than at Kingman.

As you might expect, the increased pathogen population is causing more coral disease on Christmas. Liz Dinsdale, one of the Benthics from San Diego State University, measured the percentage of diseased corals on each of the four atolls. There was a significant increase in disease incidence on Fanning, as compared with Palmyra and Kingman. *Moreover, the percentage of corals that appeared unhealthy at Christmas was more than double that which she observed at Kingman. This is particularly startling in light of the fact that there was almost no coral left on Christmas.* This pattern tells us something about the nature of those coral diseases. Infectious diseases caused by specific pathogens typically require a fairly high density of potential hosts so that the pathogens can quickly and reliably find their way to their next host. (This is one reason why epidemics tend to die out before the entire population is infected.) As a result, the prevalence of a disease caused by a specific pathogen will be greater where the potential host density is higher. The situation on Christmas is just the opposite: higher disease incidence but lower coral cover. This strongly suggests opportunistic diseases, the work of microbes that are normally present but able to cause disease only when provided with an exceptional opportunity. In this case, opportunity comes in the guise of stressed corals and excess food (DOC) fueling the microbes.

ભ § ભ

During the first Northern Line Islands Expedition in 2005, time limitations restricted the surveys on Christmas to the heavily fished reefs adjacent to London and Paris, the main villages on the atoll. In 2007, Sheila Walsh (Scripps Institution of Oceanography) led an expedition to the more remote sites. Comparing the different sites around Christmas showed that this one island has patterns on a small scale that are very similar to those observed over the four Northern Line Islands. Near London and Paris, the big fish are gone and there are more algae, more super-heterotrophic microbes, more coral disease, and less live coral. Away from the human settlements, the corals are healthy, the microbes are fewer, and the microbes are not super-heterotrophs. Even though the sharks have been essentially fished out by chumming from a distance, vampire snappers and other large apex predators are still present on these remote reefs. Ironically, the healthiest areas were the sites of nuclear testing, where lingering fear of radiation combined with rough water has deterred most fishermen.

$$\backsim \S \backsim$$

In 2009, the Fish, Benthics, and Microbes returned to the central Pacific to survey five of the most pristine coral reef ecosystems remaining on the planet. All five surround uninhabited islands in the neighboring Southern Line Islands. On all five, coral cover is very high, while seaweed and turf algae are almost impossible to find. All five islands have an inverted fish biomass pyramid with most of the biomass at the top in the form of sharks and vampire snappers. Diseased corals are rare and pathogens are undetectable in the water.

These findings confirm yet again that it is the activities of people that are causing the present decline of corals. If humans are the cause, then we can be the solution.

CHAPTER NINE

Giving Coral Reefs a Chance

Around the world, coral reefs are dying from the local impacts of overfishing and nutrient enrichment. In the future, they will also be facing intensification of the current global stressors: rising temperatures and increasing ocean acidification. Survival of the reefs—both short-term and long-term—demands the immediate reduction of fishing and nutrient enrichment, as well as the effective protection of larger regions of coral reef habitat. Coral reefs are invaluable. We have the know-how needed. Do we care enough to act?

IX

Revisiting the Reefs

The Goddess, also known as the Girlfriend, is a molecular biologist—one who studies the very, very small things that make living things work. She is on intimate terms with proteins and DNA. She knows which electron orbital belonging to which of the several hundred amino acids in which enzyme is the one that can cut a particular strand of DNA—knowledge that is important for treating cancers and developing new antibiotics. In her everyday work, the Goddess uses high-tech, Star Trek-like instruments. And strangely, she does not consider PVC tubing or duct tape to be essential laboratory equipment.

The Goddess is also fairly convinced that running around the world and SCUBA diving on scenic reefs is not real science. In fact, she has even gone so far as to distinctly imply that coral reef ecologists might just be playing. No matter how many times the Microbe explains to her the difficulties of doing research "at sea," she refers to each marine scientific expedition as a luxury cruise.

"But our cruises are work. We SCUBA dive all day. It's exhausting. And on shipboard it's self-service. We have to get our own drinks, even catch our own dinner." Of course, many people share the same delusions as the Goddess. Parents, for example, don't think that being a coral reef ecologist qualifies as having a real job. Most people don't see coral reef scientists as the rugged, highly-trained specialists that we are, uniquely qualified to do our jobs under the most grueling circumstances. It just so happens that our circumstances include going to world class dive sites and *spending long hours underwater observing the most spectacular coral reefs on the planet. Someone has to do it.*

So when the Goddess suggested to the Microbe that they return to Sipadan for their next vacation, the Microbe thought, "Here's my chance to show her how demanding coral reef science really is."

The Goddess and her personal Coral Reef Scientist and Porter had first visited the coral paradise of Sipadan in the mid-1990s. Sipadan lies within the Coral Triangle—the center of Earth's marine biodiversity—a region roughly bounded by the Philippines, Indonesia, and Papua New Guinea. Literally millions of wonderful reef creatures are to be found here, including more than 600 species of corals. Massive walls of corals plunge a thousand feet downward. Schools of giant Bumphead Parrotfish, each the size of a small car, busily crunch algae off the reef. And, of course, there are the predators: reef sharks, leopard sharks, giant trevally, and, most impressive of all, tornados of hundreds of circling Pacific barracuda.

Upon their return to Sipadan, now seven years later, the Microbe and the Goddess headed first to the Hanging Gardens— a breathtaking wall of living corals they had explored on their previous visit. This time it was horrifyingly different. The corals were dead and covered with microbial biofilms. Surveying the scene, the Microbe announced in the voice of an expert, "The coral is dead. But to be really sure I need to get a length of PVC pipe and..."

"Anyone can see that the coral is dead. Who needs PVC?"

"Well as a highly trained Coral Reef Scientist, I can tell you that in order to know that the coral is really dead we need a quadrat of PVC pipe. We have to photograph and quantitate this. Just ask the Benthics."

The Goddess mumbled something snide about ecologists and then observed, "It looks to me like the coral is being killed by disease. Probably there are too many microbes in the water."

This was too much for the Microbe, himself in the water at the time. "And just how might you know that?"

"Well, those corals that still look somewhat alive have lesions and dead patches all over them. And look at the water. We used to be able to see for hundreds of feet here, but now it's cloudy."

"Yes, but to prove that scientifically, we'll have to haul several tons of equipment out here. First we'll lay down a rope. Then we'll duct tape our underwater camera to the PVC quadrat frame and lower it into position in the water. Then we'll..."

"Yes, yes, yes. I've heard all this before. And then we'll vacuum some water off the bottom using a bilge pump, filter out the microbes, and count them. Don't you ever feel just a little silly capturing water while you're literally submerged in the stuff?"

Of course, the Goddess was right. Many of the corals were ill or dead, and the microbes were flourishing. We eventually learned that the resorts on Sipadan had been releasing their sewage into the middle of the island. Since the island is made up of nothing but porous carbonate skeletons, the sewage nutrients seeped through the ground and onto the surrounding reefs. The Goddess was not pleased to find that her Coral Gardens were dead. Having heard the story from the Microbe on more than one occasion, she recited, "The nutrients feed the algae; the algae feed the microbes; the microbes kill the corals." The Microbe sought to quietly dissociate himself from those other microbes.

Seeking to make amends for the rubble, a few years later the Microbe suggested they spend a week on a liveaboard dive boat exploring part of the Great Barrier Reef (GBR). The GBR is often touted as not only the largest coral reef system on the planet, but also the best managed. The Microbe steered the Goddess toward the coral paradise around Lizard Island that he had visited some 15 years before.

After the first dive, the Goddess was not pleased. "There are supposed to be sharks, but there aren't any."

Slightly annoyed, the Microbe asked, "How do you know that? The Fish spent literally decades learning how to count sharks and now they are the best shark counters in the world. It takes them weeks of work to determine exactly how many sharks there are per unit volume on a particular reef. And you just swim about for a couple of hours and pronounce that there aren't any sharks. How can you conclude that?"

"Because I didn't see any."

The Microbe had to admit that the Goddess was uncannily correct. There weren't enough sharks. When they dove at Cod Hole, he counted only four. On a similar dive in 1991, even without any special Fish training, the Microbe had counted 42 sharks. Where had they gone? The Microbe decided to ask the crew of the liveaboard, many of whom had been diving the reefs for a decade or more. At first he was given the standard spiel for tourists: the reefs are doing just fine...come back and visit us again...bring money. Then, finally, one of the crew fessed up. Not only is much of the "best managed" coral reef in the world actually unprotected, there is an active shark fishing industry. For the dive operators to take action on behalf of their reefs would mean calling public attention to the ongoing damage. Tourism would decline, and with it their short term profits.

The Goddess, despite her lack of training in the highly demanding field of coral reef ecology, could see that the corals are dying and that the sharks are disappearing. Others have seen this, too. They, and perhaps you as well, are asking, what can we do?

et's do a quick review of what we've learned about corals so that we can come up with effective ways to aid them. Coral reefs are in trouble worldwide. Because of *shifting baselines*, we had grossly underestimated the harmful impact of human activities on coral reefs since colonial times. Today, the most obvious sign of the accumulated damage is the loss from many reefs of the *coral holobiont*, the keystone entity that captures passing nutrients, harvests sunlight, and builds the reef structure.

The coral holobiont is an interdependent association of many types of organisms. Mix-and-match combinations of different coral animals, zooxanthellae, fungi, microbes, and viruses generate an array of slightly different holobionts that, as a group, are able to occupy a wider range of environments than any one combination could do alone. We have come to realize that several human-generated stressors are changing the microbial composition of the holobiont, notably by increasing the number of potential pathogens capable of killing the coral animal.

Some of these stressors are global, that is they affect reefs in virtually all parts of the world. Others are local, being associated with the activities of nearby humans. The most important stressor that is acting globally is the increase in atmospheric CO_2 due to our burning of fossil fuels. This has already led to *acidification* of the oceans (lowering of the average pH) and increased sea surface temperature (which stresses the coral holobionts and can even trigger *bleaching*). These CO_2-related stresses will undoubtedly worsen in the decades ahead. The foremost local stressors are *overfishing* and *nutrient enrichment*. Direct physical destruction of reefs by dredging, dynamite fishing, building on top of them, and similar activities—another local stressor—has now been curtailed in many regions, but remains a problem in some. ***While global stressors capture the headlines, local stressors are currently the main coral killers.***

Each of these stressors affects a coral reef in its own distinctive way, but all lead to a similar observable result: increased incidence of coral disease and loss of the intricate reef structure. Specific diseases, such as white band disease, have wiped out key species of coral. Epidemics have also decimated other important ecosystem members, such as the sea urchin *Diadema* whose active grazing had kept the fleshy algae in bounds on

Caribbean reefs. Overfishing and nutrient enrichment are the major local culprits. Both lead to more fleshy algae on a reef. More fleshy algae, in turn, release more DOC (dissolved organic carbon) into the reef waters. More DOC fuels a more numerous and more active microbial community, one with more pathogenic members. The result? More opportunistic and specific diseases afflicting corals and other reef organisms. This is the *DDAM model* (DOC, Disease, Algae, and Microbes) of increased coral disease. By curtailing fishing and nutrient additions, we can keep the fleshy algae in check and break the DDAM cycle that kills corals.

The stressors that act globally (increasing water temperature and acidity) also favor disease states, as exemplified by the observed reactivation of resident herpes-like viruses. Such effects may reflect either the response of the holobiont's microbial and viral communities or a breakdown of the coral's immune system.

ᘓ § ᘓ

This litany of woes may sound like the death knell for coral reefs. The situation is serious and will inevitably worsen. However, as we attempt to predict the future, it is important to remember that the coral holobiont is exceptionally adaptable. Corals and their kin have been around since the Cambrian (540 mya); some of today's coral species have existed for 55 million years. During this time, they have weathered major environmental changes. We simply do not know the extent of the holobiont's abilities to respond quickly to shifts in ocean temperature and acidity. We do know the holobiont can call upon numerous possible strategies: swapping out their zooxanthellae (adaptive bleaching), forming associations with microbes that can counter nutrient enrichment (e.g., denitrifiers that convert nitrate to nitrogen gas), and adapting physiologically to increasing acidity. Microbes are particularly well suited to function as the holobiont's first responders to stressors because they encode a far greater variety of metabolic capabilities than do the coral and the algal zooxanthellae. As a result, they are more apt to have the needed capabilities already at hand. The possibility of rapid adaptation by the holobiont, thanks to its algal, viral, and microbial partners, is a source of hope.

In the rest of this chapter we discuss the highest priority actions needed to assist coral reefs into the future. We have focused on alleviating local impacts. It is certain that the global effects of increasing atmospheric CO_2 will intensify in the decades ahead and will impact coral reefs. Policies to decrease CO_2 emissions obviously warrant support, but they have been dealt with extensively elsewhere. Furthermore, current efforts aimed at lessening CO_2 emissions and their concomitant effects would have their greatest benefit for corals perhaps a few decades from now. Meanwhile, the coral reefs require immediate help if they are to survive until then.[a]

Overfishing, nutrient enrichment, bleaching, and habitat destruction are killing corals today. Therefore, our top three priorities are:

I. Stop overfishing—especially of the large predator and herbivore species.

II. Stop nutrient additions to coral reefs.

III. Provide increased habitat protection. Provide complete protection for those coral reefs most likely to survive the global changes.

ᕦ § ᕤ

I. Stop overfishing—especially of the large predator and herbivore species.

When a coral reef has an intact fish community, the larger herbivores keep the fleshy algae well grazed. Then the energy captured by these pho-

---

[a] For those seeking suggestions for practical actions that can be taken by virtually any individual, there are numerous lists available online. For example:
  (1) from the National Oceanic and Atmospheric Administration
      25 Things You Can Do to Save Coral Reefs
      http://www.publicaffairs.noaa.gov/25list.html
  (2) from the Planetary Coral Reef Foundation
      Ten Simple Things You Can Do To Save The Coral Reefs
      http://www.pcrf.org/tensimple.html
  (3) from Project AWARE: Divers Conserving Underwater Environments
      Coral Reefs: Ten Things You Can Do to Help
      http://www.projectaware.org/assets/library/148_coralfactsandcoralreefsth.pdf

tosynthesizers fuels not only those herbivores, but also all the predators that feed on them. Fishing disrupts this pattern, resulting in more algae releasing more DOC into the water—and thus more microbes, specifically more pathogenic microbes. Curtailing fishing on coral reefs can reduce coral disease by breaking the DDAM cycle.[b]

Why are we fishing so much? Our need for food is often cited as justification. Coral reefs have the potential to be exceedingly productive, long-term food sources. Overfishing them today is a shortsighted choice that robs future generations. The very idea of "overfishing" was once thought an impossibility. The famous biologist, Thomas H. Huxley said in 1883:

> *I believe, then, that the cod fishery, the herring fishery, the pilchard fishery, the mackerel fishery, and probably all the great sea fisheries, are inexhaustible; that is to say, that nothing we do seriously affects the number of the fish. And any attempt to regulate these fisheries seems consequently, from the nature of the case, to be useless.*[c]

With this comfortable mindset, humanity readily embraced each new fishing technology as it came along. The modern industrial fishing fleet uses military equipment, such as SONAR, GPS, and live satellite feeds, to methodically hunt their prey. Thus armed, we have become extremely good at killing fish. There are now floating factories that use fishing tackle on a scale that is hard to imagine: 130 kilometer (80 mile) longlines with thousands of baited hooks, and drift nets that extend 40 km (25 miles). These monstrosities kill everything in their path.

The fishing industry is heavily over invested in its deadly capital

---

[b] To directly assess whether functionally diverse fish communities mitigated the incidence and severity of coral disease, researchers surveyed 14 sites in the Philippines. Seven sites were in well-managed protected areas with intact fish communities; the other 7 were comparable sites that were open to fishing. Their results were consistent and clear-cut: coral disease prevalence was less on reefs that were protected from fishing. Within the fished reefs, those with more diverse fish populations had less disease, suggesting that even moderate reduction in fishing pressure could have beneficial effects on coral health.

Raymundo, L.J., Halford, A.R., Maypa, A.P., and Kerr, A.M. (2009) Functionally diverse reef-fish communities ameliorate coral disease. *Proceedings of the National Academy of Sciences USA* **106**: 17067-17070.

[c] Quoted from an influential address presented by T. H. Huxley at the International Fisheries Exhibition in London in 1883.

equipment. It is estimated that there are twice as many fishing boats worldwide as would be required to bring in the maximum sustainable harvest.[d] One factor that has led to this situation is the entrenched system of government fishing subsidies. In 2006, these totaled US$30–34 billion globally. About two-thirds of the total, mostly in the form of fuel subsidies and capital for vessel construction, promotes increased fishing.[e] This needs to be changed so that the subsidies encourage the maintenance and growth of fish stocks.

Another driving force is the allure of specialty markets with high profit margins. On Christmas Island, as well as on many other coral reefs around the world, sharks are killed for money, not for food. Shark fins bring singularly high prices in Singapore, Hong Kong, and Taiwan—and the rest of the shark carcass is usually thrown overboard. So long as there is this strong demand, there will be fishermen to profit from it. Thus, one strategy to reduce finning is to undermine the market by educating the public about the cruelty and waste of shark finning, as well as about the toxic levels of mercury in shark fins. Another strategy is to endow some people with a vested interest in sharks as a profitable resource, for today and tomorrow. This implies ownership, as opposed to a *commons* open to all.

Fisheries are a classic example of the Tragedy of the Commons. This drama, with its predictable conclusion, has unfolded again and again when individuals despoil a shared resource to maximize their own short-term gain even though they know that everyone, including them, will lose in the long run. When many fishing boats are competing for the same fish, each boat seeks to maximize their catch. Why leave any fish to serve as breeding stock? Some other boat would catch and sell them. However, the Tragedy of the Commons can be averted if someone with long-term interests owns the fish. The owner then has a vested interest in protecting their resource.

One compelling example of this can be found in the traditional, community-based marine resource management systems widespread

---

[d] Sumaila, U.R., Khana, A., Watson, R., Munro, G., Zeller, D., Baron, N., and Pauly, D. (2007) The World Trade Organization and global fisheries sustainability. *Fisheries Research* **88**: 1-4.

[e] ibid.

throughout the Asia-Pacific. Only members of the local communities were allowed to harvest from the nearby fishing grounds, and they had to abide by numerous protective restrictions.[f]

This same principle of resource ownership is the basis of a promising new approach being used in fisheries management today. By 2003, this strategy, known as individual tradable quotas (ITQs), had been introduced in more than one hundred regional fisheries. For each ITQ fishery, the total allowable catch (TAC) was established based on the best estimates of the current sustainable harvest. The TAC was then divided into equal shares, twenty shares for this example, that were sold to fishermen. These 20 shareowners now have exclusive access to the fishery, with each fisherman entitled to 1/20th of the TAC. It is then in the best interests of the fishermen to lobby for a sustainable TAC. Long term, a share in a thriving fishery can be resold for a high price and thus represents real wealth for the owner. Short-term profits also improve when the TAC increases year-by-year due to this fishery management strategy.

Do ITQs work? In 2008, a hundred-plus regional fisheries, many with ten or more years of documented history, were evaluated. Not only did the probability of fishery collapse drop with ITQs, but many of the fisheries showed a marked increase in sustained catches.

If ITQs are so effective, why have they not been more widely implemented? One factor is that proof of their performance came only in 2008. And, of course, there are various political and regulatory issues in the way when a fishery spans the waters of more than one country—as many do. There can also be serious disagreement as to what the TAC should be. Determining what represents a sustainable harvest is difficult, partly because we usually lack adequate baseline data. Lastly, no matter how successful ITQs might prove to be, some people are simply opposed to them. They feel that ocean resources are common property and that everyone should have free access to them without regard to the realities of

[f] Nickerson, D. J., Maniku, M.H. (eds). Report and Proceedings of the Maldives/FAO National Workshop on Integrated Reef Resources Management in the Maldives. Male,16-20 March, 1996, Madras, BOBP, Report No. 76.
Specifically, the following two papers therein:
(1) Ruddle, K. (1996) Traditional Marine Resources Management Systems in the Asia-Pacific Region: Design Principles and Policy Options.
(2) Johannes, R. E. (1996) Traditional Management Options and Approaches for Reef Systems in Small Island Nations

fishery collapse. This sort of ignorance has to be fought head on.

The amalgamated nature of the fishing industry today means that it is not sufficient, or even correct, to point a finger exclusively at the fishermen. Most of the money ends up in other hands. Relative to the annual revenue of the commercial fishers, the processors earn almost double and the distributors more than triple.[g] In addition, out of sight of land and beyond the reach of any particular government, illegal operations abound.[h] One game that the dark side of the fishing industry plays is "hide-the-origin." A boat catches fish within the purview of a country with fishing regulations, but when questioned claims to have caught them in some other—unregulated—locale. This has become a profitable but illegal business practice, with companies running complex schemes to make it impossible to trace their catch back to the point of origin. There are documented cases where a fish has changed hands at least six times between when it was caught and when it was eaten.

To discourage such practices, it is necessary to track down the worst offenders, though they be hiding in the murky world of marine and corporate law. Small countries whose fish and other valuable marine resources are being stolen, such as the Republic of Kiribati, need international help in finding and prosecuting the thieves. Surely this is a worthy cause deserving the attention of *pro bono* lawyers, as well as lawyers looking for a lucrative venture.

Another profitable game is "mislabel-the-fish." When served up in a restaurant or filleted at your local fish market, a cheaper fish can often pass for a more expensive variety. In the USA, species designated as overfished by the National Marine Fisheries Service can masquerade as similar legal species, thus circumventing conservation efforts. To document the extent of these activities, Stephen Palumbi from Hopkins Marine Station (Stanford University) and colleagues turned to high tech DNA forensics. When they tested 77 fish fillets sold as "Pacific red snapper" at restaurants, fish markets, and groceries in Washington and California,

---

[g] http://www.hoovers.com/commercial-fishing-and-seafood-distribution/--ID__175--/free-ind-fr-profile-basic.xhtml

[h] Langewiesche, William. (2004) *The Outlaw Sea: A World of Freedom, Chaos, and Crime.* North Point Press

they found that 60% of them were some other species.[i] Most other studies of this sort have reported that more than 50% of the fish sold commercially are mislabeled, often with the intent to camouflage endangered species. This is credible, given the opportunities presented by the current state of affairs.

Clearly, this is bad for the fish. It is also stealing from the consumer, as when cheap tilapia is passed off as expensive snapper at a sushi bar. Mislabeling also prevents the increasingly aware and conservation-conscious public from making informed choices. Daniel Pauly at the University of British Columbia in Vancouver, among others, has suggested that most of the mislabeling is done by the distributors and other middlemen. The fishermen may also be complicit, being fully aware of having taken an endangered species.

To curb deceptive and illegal fishing practices, a Code of Conduct for Responsible Fisheries was adopted by the FAO (Food and Agriculture Organization of the United Nations) in 1995.[j] Compliance, unfortunately, is strictly voluntary. However, late in 2009 a new treaty with some muscle was drawn up, also under the auspices of the FAO. When this agreement goes into effect, the backdoor for illegal fishers will be slammed shut at the ports of 25 or more countries. At these ports, officials will take steps to identify, report, and deny entry to illegal vessels, thus foiling their attempts to offload their catch—a very cost effective strategy for curtailing illegal fishing.[k]

Other strategies with potential clout elicit the help of consumers concerned about the depletion of marine resources. A 2003 USA survey found that 72% of the respondents would be "more likely" to purchase seafood with an "environmentally responsible" label.[l] Such surveys are indicative of growing consumer awareness and the increasing likelihood that a segment of the market would actually shop on that basis. Money talks. One group that was listening is the Marine Stewardship Coun-

[i] Logan, C.A., Alter, S.E., Haupt, A.J., Tomalty, K., and Palumbi, S.R. (2008) An impediment to consumer choice: Overfished species are sold as Pacific red snapper. *Biological Conservation* **141**: 1591-1599.

[j] http://www.fao.org/docrep/005/v9878e/v9878e00.HTM#11

[k] http://www.fao.org/news/story/en/item/37627/icode/

[l] http://www.seafoodchoices.com/resources/documents/SCA_report_final.pdf

cil (MSC) formed ten years ago in the UK to provide an independent certification and labeling program for traceable and sustainable seafood. Responsible fishermen willingly apply for certification so their product can be marketed with the MSC stamp of approval. Their reward is access to markets willing to pay a higher price to purchase sustainable seafood of known origin from ethical businesses. Currently, 57 fisheries located around the globe are MSC certified and another 115 are under assessment.

For certification to be respected there must be no breaks in the "Chain of Custody" that starts with the licensing of the fishing vessel and includes every agent that handles a certified fish until it is purchased by the consumer. High tech tools are employed. GPS is used to pinpoint vessel locations; satellite surveillance monitors the weight of the catch by direct observation of the onboard scales; a barcode identifies and tracks each sealed carton of fish as it travels from the dock through various hands en route to the consumer. This traceability keeps illegally fished seafood out of the participating shops and restaurants, and gives consumers a choice.

This is but one of the ways that some commercial fishermen have joined in the efforts to protect fisheries. They know that their livelihoods are threatened by overfishing and they are extremely annoyed that their harvest—both present and future—is being stolen by illegal fishing. In a price competitive market, the legal fishermen are at a disadvantage. They incur higher operating costs because they buy the required licenses, maintain their boats up to standards, and pay their crews. Certification is one mechanism for rewarding legitimate operations—provided the public is committed to paying a little more for responsibly harvested seafood, as increasingly seems to be the case. The similarity between today's rising demand for sustainable seafood and the organic movement of the past couple of decades has not been missed by the business world. Anticipating increasing consumer demand, WalMart (USA) has announced that by 2012 it will purchase wild-caught seafood only from MSC-certified fisheries. Similar developments are seen in Europe.

By 2006, half of the seafood consumed by humans was produced by aquaculture, with the largest market shares seen in the USA and Europe.

Considering the rapid growth of aquaculture, by 2012 WalMart might not be purchasing much wild-caught seafood at all. Increasing our reliance on farmed fish is not a panacea, as aquaculture heavily impacts wild fisheries. Fish meal, an important component of aquafeed, is produced from wild-caught fish—in some cases, more pounds of wild-caught fish than are ultimately harvested from the fish farm. Aquaculture is already the largest consumer of fish meal; at its current growth rate, it would monopolize the world's entire fish meal production by approximately 2010.[m] The negative local impacts of aquafarms, also of concern, include the co-opting of important fish habitats, pollution, and increased disease. Remember the Gulf of Aqaba story.

There are many other legal businesses that are harmed by illegal or excessive fishing. These include those that cater to the sports fishermen, the SCUBA divers and snorkelers, and other eco-tourists. It is in their best interest to help conserve the fisheries and the reefs. Who wants to build a multimillion dollar resort only to have the nearby coral reefs—the star attraction—die? All of these people who earn their living, directly or indirectly, from the coral reefs, are potential allies in our efforts to use political and economic tools to aid the reefs.

To summarize, the fisheries of the world, including coral reefs, have been regarded as an inexhaustible commons whose wealth was there for the taking. Illegal fishing was especially profitable. Where regulations did exist, they could often be ignored with impunity. This abuse has taken a toll on coral reefs. To transform this scene into a legal industry engaged in sustainable harvesting from healthy reefs requires our taking action at numerous levels. Specifically, this calls for:

- Well-founded government regulations and strategies such as ITQs.

- Adequate enforcement of regulations.

- Redirection of government subsidies from funding increased fishing to providing incentives for ecologically sound business practices.

---

[m] http://aquaculture.noaa.gov/pdf/feeds_02hardyppt.pdf

- Consumer awareness supporting responsible fishing and eroding the market for shark fins and other key reef species such as the grazing parrotfish.

- Strong public support for such policies spearheaded by all those whose livelihood depends on coral reefs.

II. Stop nutrient additions to coral reefs.

The growth of fleshy algae on a coral reef can be regulated by top-down control (grazing), bottom-up control (limited nutrients), or both. Adding nutrients to a reef frees the algae from the bottom-up control and stimulates their growth. If grazing is brisk, some nutrient enrichment can be tolerated because the top-down control still keeps the fleshy algae in check. However, if there are relatively few grazers due to overfishing, a disease epidemic, or other factors, top-down control will be inadequate for the task. Extra nutrients arrive on a reef as side effects of poorly-managed human activities, particularly agriculture, deforestation, land clearing, and sewage dumping. As with overfishing, reducing these local impacts calls for both the enforcement of government regulations and for innovative market-driven strategies. For example, the runoff of agricultural fertilizers represents tens of millions of dollars in lost nitrogen and phosphate compounds every year. Practices that reduce runoff and/or recover nutrients and return them to the soil could be money-makers for farmers and for nutrient recovery operations. Coral reefs would benefit directly from the reduced nutrient loading and indirectly in that reduced fertilizer production would reduce greenhouse gas emissions. Some well-placed legislation and development funds could stimulate technological innovation in these directions.

Economics also influences where our sewage goes. Sewage treatment costs money in the short-term but can bring long-term profits. The ongoing economic value of a lively reef is far greater than the cost of building and operating treatment facilities. However, local governments often need funds from the outside to cover the initial costs. When viewed from a distance, such programs may seem to be straightforward and destined to be successful. However, the reality can be more complex. Consider, for example, Christmas and Fanning in the Northern Line Islands. The coral reefs are their only tourist attractions. Money invested in sewage treatment by the Republic of Kiribati could be recouped through tax revenues from a thriving tourism industry serving SCUBA-diving visitors and sports fisherman. The accompanying reduction in pathogens in the reef waters would also improve the health of the people of Kiribati. One other ingredient is needed: education. Installing toilets that feed to a sewage treatment plant can reduce reef nutrients only if the people actually use them.

Likewise, home septic systems can adequately treat sewage only if used properly. In some rural areas of the USA, homeowners are required to install an approved septic system. The household waste is then carried by pipe to a closed septic tank for digestion, and from there the effluent flows to a drain field where it percolates into the soil. When the septic tank is too small for the household or its activity is sluggish due to cold temperatures or to toxic chemicals sent casually down the drain, the tank fills up with solids. The homeowner can then pay for a honey truck to come pump out the contents and haul them away. The more economical solution is to knock a hole in the tank, which is said to happen in the Florida Keys and elsewhere. Even in a developed country, more education is needed.

III. Provide increased habitat protection. Provide complete protection for those coral reefs most likely to survive the global changes.

We are decreasing the resilience of coral reefs today by overfishing, by nutrient addition, and by habitat fragmentation. When visualizing the alternative stable states of an ecosystem as valleys in a landscape (see pages 92-94), decreasing resilience changes the size and shape of the valleys making it easier for a perturbation to shift the ecosystem from one stable state to another. A resilient coral reef can tolerate stronger perturbations without shifting from a coral-dominated to an algae-dominated state.

Reef resilience, as appraised by David Obura (CORDIO: Coral Reef Degradation in the Indian Ocean), is high for the pristine reefs within the Line Islands, the Phoenix Islands, and other remote Pacific Islands. High resilience does not exempt these reefs from environmental stressors, but it can influence the outcome. When corals around the Phoenix Islands experienced strong temperature stress (16 degree heating weeks) in 2002-2003, at least two thirds of them died. However, these resilient reefs did not shift to an algae-dominated state even though coral cover was dramatically reduced. Instead of an overgrowth of fleshy algae, the bottom is now covered by abundant CCA and many young corals—both of which bode well for reef recovery.

Thus, our first step in aiding coral reefs today is to increase their resilience by protecting them from overfishing, nutrient enrichment, and habitat fragmentation. This is also the optimal strategy to help them cope with the inevitable increased levels of atmospheric CO_2 that lie ahead. We can already predict where increasing acidification and sea surface temperatures will be the *least* severe. The reefs in these survival zones will have the best chance for long-term survival and therefore should be given top priority for assistance. Coral reefs in these survival zones should be *completely* protected from local stressors.

Already some very important steps have been taken in this direction. In 2008, the Republic of Kiribati designated eight uninhabited atolls and two submerged reef systems in the Phoenix Islands as a marine protected area (MPA)—the world's largest MPA at that time (410,500 square kilometers, 158,500 square miles, representing 12% of Kiribati's watery domain). Notably, protection also applied to the Exclusive Economic Zone that extends 200 miles out from shore and includes nearby seamounts where many fish congregate. The proposed Phase II and Phase III would add the uninhabited islands within the Southern Line Islands and selected parts of the Northern Line Islands, respectively. This impressive initiative by a poor country needs to be monetarily rewarded by the richer nations and individuals of the world. Being isolated and remote, these islands are highly vulnerable to resource extraction by illegal foreign vessels, including shark fin fishers and live reef fish collectors for the aquarium trade.

Similarly, in 2009 leaders of six countries bordering on the Coral Triangle—Indonesia, Malaysia, Papua New Guinea, the Philippines, Solomon Islands, and Timor Leste—launched the Coral Triangle Initiative on Coral Reefs, Fisheries, and Food Security.[n] This region of extraordinary biodiversity is home to 36% of the world's coral reefs, more than three-quarters of all known coral species, and an enormous number of other marine organisms. It is a treasure economically, as well, directly supporting millions of people in those bordering countries and generating billions of dollars of income globally. Like coral reefs elsewhere, these reefs are threatened by overfishing, illegal fishing, unsustainable coastal development, and pollution. Sharing a common goal—to protect the well-being and livelihood of their people for generations to come—these leaders came together and adopted a comprehensive ten-year ocean conservation plan for their region.

Other independent Pacific island nations have made commitments to reef conservation by creating a network of reef MPAs. All of these efforts are literally our front lines in the battle to save coral reefs. They deserve our support, including monetary assistance. The richer countries also have an obligation to protect the coral reefs within their own territories. Although Australia's Great Barrier Reef Marine Park is often lauded as an example of large-scale reef protection, until 2004 only 4.6% of the park area was off limits for fishing. In that year, the "no take area" was increased to one third of the park, now covering an area of 114,800 square kilometers (44,333 square miles). However, the other two-thirds remain open to shark fishing, harvesting for the live reef fish trade that supplies restaurants world-wide, and other high-impact practices.

Recently, the USA also made some significant advances. In 2006, President George W. Bush's administration created a large MPA in the undeveloped Northwestern Hawaiian Islands: the Papahānaumokuākea Marine National Monument, covering over 360,000 square kilometers (140,000 square miles) of ocean. Then, in 2009, three additional marine national monuments[o] covering 505,050 square kilometers (195,000 square miles) were designated that encompass USA holdings in the Line Islands (including Kingman Reef and Palmyra Atoll), American Samoa,

[n] http://www.nature.org/pressroom/press/press4034.html
[o] These three new monuments are the Marianas Marine National Monument, the Pacific Remote Islands National Monument, and the Rose Atoll National Monument.

and the northern Marianas. This was a significant step. Now we need to muster the political will to fully protect these national treasures. Fishing is prohibited in only about 60% of the area encompassed by these new monuments; expansion of the "no take area" to 100% and provisions for adequate enforcement should be next.

A reef can be designated as "protected" with the stroke of a pen; effective enforcement of that status is not so easy. A protected area is not protected unless the vessels illegally entering and/or removing fish or other resources are detected and intercepted. Due to the large expanses and remote locations involved, illegal fishers knew the odds were small that they would be caught. Today, we can increase those odds and actually enforce regulations through increased local surveillance and high-tech remote sensing (e.g., satellite and aerial surveillance, and passive sonar monitoring). A potentially effective strategy for curtailing illegal fishing is in the offing as 25 countries agree to close their ports to illegal fishers attempting to offload their catch.[p]

Comprehensive reef protection also entails continuous monitoring of their condition. Sea surface temperature can be monitored via weather satellites, thus providing advance warning of a major bleaching event. By measuring the light reflected from the reefs, other types of satellite imagery detect some gross changes to the reef, such as the loss of coral cover or coral bleaching. On-site monitoring is even more important. *In situ* data loggers that measure water temperature and pH, microbial activity, and other key environmental parameters provide data that can help us identify the factors that drive the shift of coral reef ecosystems from a coral-dominated to an algae-dominated state. One regional monitoring network already in place is SEAKEYS, organized by the Florida Institute of Oceanography and encompassing a network of seven monitoring stations spanning the Florida Keys and the Dry Tortugas.[q] Hourly readings of atmospheric and oceanographic parameters are transmitted to a Geostationary Orbiting Environmental Satellite and made available in near real-time on the Internet.

On a global scale, observations made by numerous agencies and gov-

---

[p] http://www.fao.org/news/story/en/item/37627/icode/

[q] http://www.keysmarinelab.org/seakeys.htm

ernments are being coordinated by GOOS (the Global Ocean Observing System), an ambitious, cooperative international effort to monitor the entire expanse of the world's oceans.[r] The current focus is primarily aimed toward furthering immediate human interests (e.g., weather prediction, protecting life and property on coasts and at sea, and improved ocean resource management). Temperature and salinity—key to weather prediction— are measured from space and also *in situ*; water level is monitored as part of a tsunami early warning system. Coral reefs have not been given the attention they warrant, but Coastal GOOS has the potential to provide the needed monitoring. What is called for here is *in situ* measurement of temperature, salinity, pH, light level, and other parameters at reef depths, accompanied by automatic uploading of the data via satellite to the Internet where it can be accessed by both researchers and the public using applications such as Google Ocean. GOOS, SEAKEYS, and similar programs merit expansion and adequate financial backing.[s]

ↄ § ↄ

Since you are reading this book, chances are good that you already care about coral reefs. Perhaps your livelihood depends on them in one way or another. Possibly you have seen them firsthand or are dreaming of such a trip. Likely you have felt the excitement that they arouse in many different ways. Aesthetically—coral reefs are some of the most beautiful things in the Universe. They should be valued as highly as works of art. Intellectually—the more we learn about them, the more enthralled we become. Spiritually—we are captivated whenever we briefly sojourn through their world, either in liquid reality or in our imagination. Proof? Witness a two year-old child spellbound for hours by *Finding Nemo*.

This is not just our experience; ask anyone who has entered the water world by snorkel or SCUBA. They all vividly remember their first time on a coral reef. They may excitedly tell you about the countless reef fish, and how they dart or drift about, some with comical expressions or

---

[r] http://www.ioc-goos.org/component/option,com_frontpage/Itemid,1/

[s] In contrast to the data collection provided by these continuous monitoring programs, the National Oceanic and Atmospheric Administration (NOAA) sends a vessel to remote USA-owned reefs once every two years, and even this small effort is beset with funding obstacles.

dressed in designer attire of unimaginable hues. Or they may recount their adrenaline rush when a school of barracuda encircled them or a pod of dolphins came to investigate their visitor. There aren't many human experiences that come close.

Coral reef biodiversity is amazing and essentially still uncharacterized. Even though they occupy less than 1% of the Earth's sea floor, coral reefs are home to an estimated 9,000,000 species, including about one quarter of all marine fish species. One might suppose their value would be obvious. Nevertheless, when the discussion turns to the value of coral reefs and the need for their protection, "value" often has to be expressed in dollars.

So how do you calculate the value of the world's coral reefs? Economists try to estimate the "value" to human welfare provided by reef "ecosystem services." The first service that comes to mind is the production of harvestable seafood. However, today the greatest share of the revenue from reefs comes from tourism and recreation. Protection of the nearby coastline is also high on the list. Not included in these totals, but of potentially immense value, are the myriad natural products synthesized by diverse reef organisms, any one of which could be found to be a potent drug for treating human disease.

What does all of this add up to? Estimates of the total annual services from coral reefs vary from a very conservative US$30 billion[t] to a more generous estimate of US$172 billion.[u] Even the lowest estimates represent large sums. Pornography, often cited as a major industry, brings in less than $10 billion per year in the USA. Even Hollywood, including its associated worldwide revenue streams, tallied only about $45 billion in 2004.[v] Coral reefs, although of small expanse relative to the whole of the world's oceans, are extremely valuable.

Since it is difficult to comprehend such large numbers, let's look instead at a small portion of a coral reef, one square kilometer (one-third of a square mile). In 2006, the United Nations Environment Program

---

[t] http://www.icriforum.org/library/Economic_values_global%20compilation.pdf
[u] http://www.scientificamerican.com/blog/post.cfm?id=how-much-are-coral-ecosystems-worth-2009-10-22
[v] http://www.edwardjayepstein.com/mpa2004.htm

valued a typical square kilometer (247 acres) of reef at US$100,000 to US$600,000 per year, increasing that to US$1 million in parts of Indonesia and the Caribbean where tourism is the main use.[w] Income from reef-related tourism on the GBR tallies in the billions per year, and likewise in the Caribbean. These numbers translate into the livelihoods of many millions of people. Does it make economic sense to protect coral reefs? That same United Nations Environment Program report found that the annual management costs for marine protected areas is only US$775 per square kilometer (247 acres).

Despite their obvious appeal, and despite their real economic return on investment, it remains a major challenge to raise the funds needed to protect coral reefs. It takes money to establish and manage MPAs and ITQs, to assist communities during transition periods as reef management policies are put in place, to build sewage treatment plants, to install and maintain a reef monitoring network, and to fund the research. Some of this money can come from governments that are willing to invest now for greater future returns. Some can come from business entrepreneurs that see opportunities for profit in recycling nutrients, certifying seafood, or designing innovative sewage treatment facilities. Some can come from individuals, the sorts of people who donate hundreds of millions of dollars for the establishment of a new public art museum. Why not protect coral reefs as well as we protect other valuable works of art? If you have such resources, consider endowing a coral reef. If you have less, consider joining with others by donating to one of the effective reef conservation groups. Or combine financial support with a memorable first-hand coral reef experience by signing up for an eco-friendly tour of one of the premier coral reefs of the world. Ecotourism puts money in the hands of people with a vested interest in preserving the coral reefs. But be sure to do some homework before you set out, and while there take care not to knock over any corals. Careless tourism can be worse than no tourism.

Finally, open your imagination to other solutions. Why not move the coral reefs? Even though it is more science fiction than science, imagine

---

[w] News release by the United Nations Environment Programme (UNEP) in 2006. http://www.unep.org/Documents.Multilingual/Default.asp?DocumentID=466&ArticleID=511 2&l=en

a massive floating platform that sails itself using wind and solar energy. Seed it with coral nubbins and other reef organisms. Post a *No Fishing* sign. Then, as the effects of increasing atmospheric CO_2 worsen, guide it to the most hospitable areas of the ocean, thus avoiding the most acidic regions and the warmer waters that would bleach the corals. Allow some reef-conscious entrepreneurs to conduct eco-tours so that people remember those marvelous constructs that once flourished naturally offshore of many islands and coastal lands. Thus sustained, we may remain dedicated to their survival and to their ultimate repatriation when the climatic upheavals have played out.

Granted, there might not be any floating reefs in the near future, but the take home message is this: Each one of us can help save coral reefs. Be creative and explore unique options. The coral reefs of the world are worth it.

EPILOGUE

Going Home

The Fish had a major instrumentation error.

"What do you mean, 'The paper won't work'? How can paper not work?"

But it was true, pencil marks just rubbed off the new paper. This was a new, high-tech underwater paper that, evidently, had not actually been tested underwater. This was an epic setback for the Fish, and it prompted much sympathy from the other scientists.

"Did you hear the Fish can't get their paper to work?" chortled one Microbe, tears of laughter running down her face. "That's got to be the funniest thing ever."

While the Benthics and Microbes continued to drink beer and laugh their heads off, the Fish schooled to come up with a solution. For this traditional conferencing, the eldest Gruff and Grouchy Fish takes a position in the center, surrounded by a concentric circle of F3ish, while on the periphery, where they can be more easily picked off by predators, the Minnows excitedly dart about. "Can't we use our computers underwater and just enter the data directly into Excel?" suggested a Minnow. One of the wise, elder F3ish nodded sagely. "Probably, if we enclose it in a plastic bag to keep out the water." Pushing their technical know-how to the limits, the Fish decided that a clear plastic bag would be best because they needed to see the keyboard.

As the Fish prepared to test out their idea with a Minnow's laptop and large Ziploc® bag, one of the Microbes helpfully suggested that since they might run out of power at a critical point while underwater, it would be best to run an extension cord out to the dive site. This was greeted by the Fish as an excellent idea. "Thank you! We didn't know you cared."

"No worries, just hold onto this end of the cord. I'll plug it in, and then you step into the seawater-filled boat," purred the Microbe.

Finally, one of the Benthics couldn't take it anymore. "Cut it out! You might destroy a valuable computer." The Benthics aren't really that much fun.

Eventually, after much testing, the Fish determined that if they filed down their pencil stubs just right and wrote very lightly, the malfunctioning paper would work 34% of the time. This was deemed good enough. Surprisingly, it had no effect on their resulting data.

Then the Microbes ran into a similar technical issue. "Ummm...I think we need to count something underwater."

"We just capture water underwater. We don't count underwater. That's for the Fish and the Benthics."

"Yes, but this time we need to count the number of coral-algal interaction zones. We need an expert." After the shark sampling incident, the Microbe approached a Fish with some trepidation. "Can you teach us how to use underwater paper and pencils?"

"Finally, you're going to see how hard it is!" triumphed a Fish. "It's extremely complicated and mastery requires extensive training, but I can give you a crash course. First, you need to get two large rubber bands. Then you use the rubber bands to hold the paper to the dive slate, one at the top and one at the bottom, like so. And then you write on the paper with the specially filed pencil. Did I tell you I got a B+ in this class at graduate school?" Turns out, Bio678L, Underwater Paper Usage, is another weed-out class for Minnows.

After being schooled in the Fish's trade secrets, the whole crowd—Fish, Benthics, and Microbes—headed out to the reef.

Time was short. The cruise was ending. This was the last dive, and everyone was trying to get that one remaining essential bit of data.

The reef was spectacular. Millions of rainbow-colored anthais hovered peacefully over giant coral heads that were hundreds of years old. Transect lines were laid down, and the vampire snappers promptly came in and started to bite them. Sharks cruised slowly past iridescent coralline algae. Banded shrimp sat on a coral outcropping, waving their antennae, soliciting customers for their cleaning station. Data was collected and then everyone's air pressure started to sink towards 500 psi. Time to go home.

The one survival law that even the Minnows obey is called the safety stop. When you SCUBA dive a lot, nitrogen gas builds up in your tissues and blood. If you come up too fast, the gas can expand and cause 'the bends.' 'The bends' are extremely painful, can kill you, and even ruin your whole day. If you hang out ten feet below the boat for five minutes on the way up, much of the nitrogen can escape.

As the Fish, Microbes, and Benthics hung quietly in the water beneath the dive boat, a gigantic Napoleon wrasse swam over to look at these strange intruders. Napoleon wrasse are possibly the most lovely and the most comical fish ever birthed by Natural Selection. Overall length is an impressive six feet or more. They don't use their tails to swim. Instead the massive wrasses pull themselves along using their ridiculously small pectoral fins. Somehow they make this look forlorn and majestic at the same time, and everyone stifles their impulse to laugh.

And then there are their eyes, which Napoleon wrasse can roll around independently. They've always got one eye watching you while the other scans the underwater landscape. A big wrasse like this one is likely thirty years old—older than many Minnows—and his goggle-eyed stare seems uncannily wise. He and his kind have been puttering around these reefs for millions of years. Everyone watches the wrasse. He watches us back.

And we are reminded why we need coral reefs.

APPENDIX A

Identification of Microbes Using 16S rDNA

The Basics: Recall that the DNA (deoxyribonucleic acid) molecules inside every cell encode the inherited genetic instructions for how to build and run that particular type of organism. This information is encoded in the sequence of four different subunits (bases) that are chained together to make up the long DNA molecules. Two molecules of DNA, termed *strands*, are twisted around each other to form the celebrated double helix. The two strands are connected by hydrogen bonds (chemical bonds that are weaker than covalent bonds) that form between each base in one strand and the base opposite it in the other, thus making the double helix a stable structure. These hydrogen bonds are discriminating. In shorthand, the four bases in DNA are A, T, G, and C; the pairs that hydrogen bond together are A–T and G–C. For this base pairing to occur at each of the thousands or millions of bases along the helix, the sequences of the two strands must be complementary, i.e., every A in one strand is opposite a T in the other, every G opposite a C. This base pairing is also central to the way that information is conveyed when the DNA is replicated and when it directs cellular activities.

DNA is primarily an information storage device. Working copies of some of this information are made by synthesizing a strand of a similar molecule, ribonucleic acid or RNA, that is complementary to a segment of the long DNA. Such a DNA segment is typically referred to as a gene. This process is called *transcription* because it transcribes the information encoded in the DNA sequence into the sequence of the new RNA strand.

Information is encoded in much the same manner in RNA as in DNA, the main difference being that RNA uses the base U instead of T. Thus, the complementary base pairs in RNA are G–C and A–U.

Most transcribed RNAs function in the synthesis of proteins. Proteins are long chain polymers built from twenty amino acids ordered in a specific linear sequence. Their functions are diverse. Many are structural molecules that constitute the architecture of the cell; others are enzymes that catalyze metabolic activities. One group of RNAs, the messenger RNAs or mRNAs, specify the precise sequence of those amino acids for each specific protein. This information is encoded in the base sequence of the mRNA.

Other RNAs, the ribosomal RNAs or rRNAs, are structural components of the ribosomes—complex cellular machines that assemble amino acids into proteins according to the sequence encoded by the mRNA. Each ribosome is composed of a large and a small subunit, each subunit in turn being a complex assemblage of specific ribosomal proteins and rRNAs. Since every cell synthesizes proteins, every cell must have ribosomes, and thus every cell must have the genes that encode the ribosomal proteins and rRNAs. For our microbial story, the most important of these ribosomal components is the 16S rRNA, one of the rRNA molecules found in the small subunit.

The Evolution and Structure of 16S rDNA: The gene that encodes the 16S rRNA is known as 16S rDNA. Evolutionarily speaking, it is ancient. It was present in a common ancestor that gave rise to all of today's cells. Over billions of years, it has evolved differently in different lineages so that today the base sequence of the 16S rDNA of each microbial species is distinctive. This alone would make it useful in identifying the microbes associated with corals, but it has also been extraordinarily fruitful for inferring their evolutionary relationships. The reasons for this are tied to the structure of the 16S rRNA.

Unlike the double-stranded DNA that forms a stable double helix along its entire length, cellular RNAs, including 16S rRNAs, are single-stranded. However, given a suitable base sequence, a single-stranded rRNA can fold back on itself to form a double-stranded region stabilized by hydrogen bonds. For this to occur, there must be strings of bases with complementary sequences located in different regions of the same molecule. When these regions find each other, hydrogen bonds form between them. This typically creates a short *stem* of double helix with a single-stranded *loop* at one end. There are many such stem loops within each rRNA molecule that, combined, dictate a specific three-dimensional shape for the molecule—a shape that is essential for ribosome function. Mutations in the 16S rDNA that alter this 16S rRNA structure impair protein synthesis and result in the death of the mutant cell.

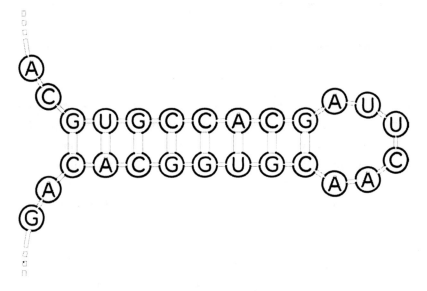

One common type of mutation is the substitution of one base for another in the DNA. If this changes a base within the stem region, the new base will not be complementary to its partner—unless its partner mutates to the complementary base at exactly the same time, an extremely unlikely event. If the bases in the stem region are not complementary, stabilizing hydrogen bonds cannot form and the stem-loop may unravel. Without the stem-loop, the ribosome cannot function, the cell dies, and the mutation is eliminated from the population. This strong selection pressure means that the sequences in these rRNA stem regions have rarely changed, even over long evolutionary time periods. It is these *conserved regions* that provide scientists with "handles" by which to recognize and take hold of the 16S rDNA genes, even in a complex mishmash of total cellular DNA from many different organisms.

In contrast, mutations in loops or other single-stranded regions can more often be tolerated because they interfere less with ribosome function. Mutational changes to these *variable regions* of the 16S rDNA can be passed on to succeeding generations. Slowly, over evolutionary time periods, these variable regions diverge in different lineages. Their differences enable us to distinguish even closely related organisms from each other.

Identifying Unculturable Microbes: The existence of both conserved and variable regions within every 16S rDNA is a piece of extremely good fortune. This has enabled microbiologists to identify the microbes present in an environment without having to culture them. Suppose you collect the microbes from the surface mucus layer of a coral and extract the DNA from the entire sample. There is a technique you can use to make many copies of all of the 16S rDNA genes—and only those genes. By sequencing those 16S rDNAs and comparing their sequences to those from all known microbes you can determine what groups of microbes are in the mucus and estimate how many different types are there.

The technique used to selectively amplify the 16S rDNA genes in a mixture of microbial DNAs is PCR, the DNA polymerase chain reaction. DNA polymerase is an enzyme that, given a length of single-stranded DNA and the right conditions, accurately synthesizes the complementary DNA strand. For it to copy 16S rDNA genes, "bookends" must be placed at both ends of the gene. Conveniently, there are conserved regions near both ends whose sequences are the same in almost all Bacteria as well as some Archaea. We can use those "handles" to create the bookends.

PCR is carried out as a repetitive process with the copies produced in one round used as additional templates for the next round of synthesis. This exponential chain reaction can yield millions of copies of the original 16S rDNA genes from bookend to bookend. These genes typically include some variable regions, as well, which can be used to identify the microbes in the sample.

The 16S rDNA Revolution: Studies using 16S rDNA revolutionized our understanding of the evolutionary relationships among all organisms. Historically, this revolution began in the mid-1980s when Carl Woese re-evaluated the kinships among known microbes and eukaryotes by comparing their ribosomal RNA genes. The Tree of Life has not been the same since. Pre-Woese, all life forms had been divided into Kingdoms, the number ranging from two to as many as five at various times, some with comfortably familiar names such as plants, animals, and fungi. Woese demonstrated that, instead, all life forms belonged to one of three main divisions, termed Domains: the Archaea, the Bacteria, and the Eukarya (the latter including all the plants, animals, fungi, and many others). This work was instrumental in introducing the current molecular-based system of microbial taxonomy (classification) and in shedding more light on the evolutionary history of all organisms.

Norm Pace took this a step further, adapting the Woesian methods for rDNA taxonomy to the study of currently unculturable microbes—the vast majority of microbes. Pace is an expert in structural and enzymatic RNAs. Coming from this background, he used PCR to search the DNA extracted from environmental samples for genes encoding ribosomal RNAs from previously unknown organisms. This approach was wildly successful and is the main way that microbes are identified today.

Metagenomics: The 16S rDNA is only one of thousands of genes in each genome. However, unlike the 16S rDNA, most of these genes do not have conserved "bookend" regions that could make them easily recoverable via PCR. So a number of groups began isolating random DNA fragments from the environment without the PCR step. They would then try to determine the function encoded by that unknown DNA. Diversa, a company in San Diego, was one of the main proponents of this approach, which eventually became known as metagenomics (named by Jo Handelsman).

It was Ed Delong, a former member of Norm Pace's lab, who demonstrated the potential of metagenomics for making serendipitous discoveries. His research group at the Monterey Bay Aquarium Research Institute used metagenomic techniques to sample and analyze longer lengths of DNA from marine environments. These pieces of DNA were long enough that they could include a microbial 16S rDNA gene plus some neighboring genes. When they found a long sequence that contained a 16S rDNA gene, that gene was used to identify the source microbe. They also sequenced the adjacent DNA. Sometimes those neighboring genes could be identified by comparison against a database of known gene sequences. Delong and his colleague Oded Beja got lucky. Using this strategy, they found the first bacterial gene known to encode a light-harvesting protein of the rhodopsin family. This was exciting because it represented a novel system for capturing sun energy in the ocean that is evolutionarily distinct from photosynthesis and that had been previously unknown. This important finding gave a well-deserved boost to the young field of metagenomics.

Metagenomics has created many research opportunities, but taking advantage of them has required new methods of analysis. When Mya Breitbart, Anca Segall, and I (all researchers at San Diego State University) initiated the use of metagenomics for studying uncultured marine viruses in 2003, we had to develop new computational methods to analyze the data. Viruses lack a shared gene comparable to the 16S rDNA. So instead we took a "shotgun" approach, simply sequencing random pieces of DNA recovered from the viruses collected from samples of seawater—the method now known as shotgun metagenomic analysis. For the new mathematical and bioinformatics tools we needed, we turned to the SDSU Math Group: Peter Salamon, Ben Felts, James Nulton, Joe Mahaffy, and Bjarne Andresen. Among those tools was a method they worked out for determining the number of species present in a metagenomic sample based on the shotgun sequence data.

We also pioneered the use of next generation pyrosequencing as performed by 454 Life Sciences (a biotechnology company specializing in high-throughput DNA sequencing) to sequence the DNA extracted from uncultured microbial and viral communities. Such a huge amount of data is generated that, once again, new bioinformatics techniques had to be devised for the analysis. In response, in 2008 Rob Edwards created the MG-RAST (MetaGenome Rapid Annotation using Subsystem Technology) server to annotate metagenomic sequences. To *annotate* a metagenome means to identify genes present in the sequences and then to predict their function by comparing them to known genes in a database. In this case, gene sequences in a metagenomics dataset are compared to those in the SEED, a database of protein-encoding genes that have been classified into groups (subsystems) that encode proteins with related metabolic functions. The microbial and viral communities sampled by the Northern Line Islands Expedition were analyzed by this powerful combination of 454 pyrosequencing and MG-RAST annotation. Interpretation of the massive shotgun metagenomic datasets from the Northern Line Islands required novel statistical techniques. Enter Liz Dinsdale, one of the Benthics skilled in identifying corals and their diseases. During the expedition she became interested in the microbiology and subsequently applied her statistical expertise to develop a new method for the analysis of the Northern Line Islands metagenomics data.

APPENDIX B

Coral Defenses

Corals have diverse strategies for protecting themselves from pathogens and parasites. Physical defenses include barriers of aragonite constructed by the coral to thwart attack by the endolithic fungi trying to bore through their skeletons. The conspicuous surface layer of mucus also acts as a physical barrier. Both the amount and the composition of the mucus are species-specific and can vary in response to environmental stresses. Mucus is not just goo, but a complex, dynamic, essential product of the holobiont. When the corals bleach, mucus production drops and the coral is more susceptible to disease. The mucus also protects the coral from UV irradiation and desiccation; its sloughing helps clear sediment from the surface and hampers microbial colonization.

The mucus layer of a healthy coral is home to an abundant population of coral-specific microbes that may prevent colonization by outsiders, including potential pathogens. Notably, these microbial partners are generally more resistant to the antibacterials secreted routinely by the coral than are the water column microbes. When injured, the coral rapidly ups the antibiotic dosage in the mucus and releases some into the surrounding water. Specific compounds have not been identified, but they include potent, broad-spectrum antibiotics. And the coral responds quickly—releasing these compounds within 15 seconds following a simulated predator bite.

Eventually, the physical barriers will fail and some microbes will invade the coral tissue. At this point, the coral's immune system will kick in. However, we know almost nothing about the details of this response.

Although corals belong to a group of the most ancient animals on the planet, they appear to have an innate immune system that can distinguish between self and non-self, and destroy the latter. In corals, there is some evidence of specialized cells, called phagocytes, that recognize, engulf, and destroy microbial invaders. ROS (reactive oxygen species) produced as by-products of normal cellular metabolism can also function as "antibiotics." Their effectiveness against a specific invading pathogen has been demonstrated in the Mediterranean coral *Oculina patagonica.*

Corals do not appear to have an adaptive immune system. There is no evidence of immunological "memory;" a coral exposed to a pathogen does not mount a more vigorous response the next time it encounters the same pathogen. While we might think the coral immune system "primitive" in this respect, we need to remember that corals are possibly immortal. The immune system they have works so well that a coral holobiont can exist for a thousand years or more. We need to understand how they accomplish this amazing feat, for our health as well as theirs.

Geffen, Y., Ron, E.Z., and Rosenberg, E. (2009) Regulation of release of antibacterials from stressed scleractinian corals. *FEMS Microbiology Letters* 295: 103-109.

Hemmrich, G., Miller, D.J., and Bosch, T.C.G. (2007) The evolution of immunity: a low-life perspective. *Trends in Immunology* 28: 449-454.

Koh, E.G.L. (1997) Do scleractinian corals engage in chemical warfare against microbes? *Journal of Chemical Ecology* 23: 379-398.

Krediet, C.J., Ritchie, K.B., Cohen, M., Lipp, E.K., Sutherland, K.P., and Teplitski, M. (2009) Utilization of mucus from the coral *Acropora palmata* by the pathogen *Serratia marcescens* and by environmental and coral commensal bacteria. *Applied and Environmental Microbiology* 75: 3851-3858.

Ritchie, K.B. (2006) Regulation of microbial populations by coral surface mucus and mucus-associated bacteria. *Marine Ecology Progress Series* 322: 1-14.

Rosenberg, E., Koren, O., Reshef, L., Efrony, R., and Zilber-Rosenberg, I. (2007) The role of microorganisms in coral health, disease and evolution. *Nature Reviews Microbiology* 5: 355-362.

Rypien, K.L., Ward, J.R., and Azam, F. (2009) Antagonistic interactions among coral-associated bacteria. *Environmental Microbiology* 12: 28-39.

APPENDIX C

Reductionism versus Holism

Don't be distressed if you have a hard time seeing how plotting the number of sharks (see pages 91-92) could actually represent the number of microbes, sponges, viruses, etc. Because our brains are not good at visualizing more than three-dimensions, we collapse highly complicated circumstances into a small (and thus conceivable) number of proxies. We chose to dedicate one axis of our graph to showing shark abundance because it is our biased view that sharks are particularly important members of the ecosystem. In fact, any time a scientist makes a hypothesis, they narrow their world down to a small number of variables that they think are the most important. Choosing the right variables is key. This reductionist hypothesis-driven science has been amazingly powerful for investigating our world.

An alternative and more holistic approach that is becoming the choice of many biologists is *systems biology*. In practice, a systems biologist assembles massive datasets that include measurements for every variable conceivably associated with the system of interest, e.g., a coral reef. Then, using a lot of computer-driven number crunching, all of the data is plotted in a mathematical hyperspace. A hyperspace is not limited to three dimensions. Thus we no longer have to condense all of the biological variables into a single axis (dimension) plotting shark abundance. There can be separate dimensions for sharks, microbes, sponges, viruses, small fish, etc., and likewise for every measurable geographical and chemical variable. For each of these dimensions, there will be a range of values that have been actually measured for coral reefs, e.g., between 23.4° S and 23.4° N for latitude. This range of values defines the region in hyperspace where coral reefs are found. This seems simple enough, but latitude co-varies with other variables such as sea surface temperature, coral species, and nutrient level. All of the combinations of all of the measured variables found on coral reefs together define the total "volume" occupied by coral reefs in this multidimensional hyperspace.

We humans cannot envision such a hyperspace. Computers, on the other hand, have no problem analyzing a hyperspace and locating the range of values in each dimension that can support a thriving coral reef. A computer can determine which variables are the hallmarks of coral reefs. These will likely be the most informative ones when asking what causes a coral-dominated reef to shift to one overgrown by algae. This kind of analysis is not biased by previous training, expectations, etc.

FURTHER READING

BOOKS

Birkeland, C. (1997) *Life and Death Of Coral Reefs*. Springer.

Corfield, R. (2003) *The Silent Landscape: The Scientific Voyage of HMS Challenger*. Joseph Henry Press.

Darwin, C. (2009) *The Structure and Distribution of Coral Reefs*. General Books LLC.

Dobbs, D. (2005) *Reef Madness: Charles Darwin, Alexander Agassiz, and the Meaning of Coral*. Pantheon.

Langewiesche, W. (2004) *The Outlaw Sea: A World of Freedom, Chaos, and Crime*. North Point Press.

Pollan, M. (2006) *The Omnivore's Dilemma: A Natural History of Four Meals*. Penguin Press.

Riegl, B. and Dodge, R. E., eds. (2008) *Coral Reefs of the US*. Springer.

Steinbeck, J. (1951). *The Log from the "Sea of Cortez."* Viking Press.

Veron, J.E.N. (2000) *Corals of the World*. Sea Challengers.

INTERNET RESOURCES

Coral Reef Systems
http://coralreefsystems.org
http://phage.sdsu.edu

Expedition Blogs from the Line Islands and Phoenix Islands
from the **Scripps Institution of Oceanography**
http://www.sio.ucsd.edu/lineislands/index.cfm

from the **Conservation International**
http://blog.conservation.org/?s=phoenix

from the **National Geographic**
http://ocean.nationalgeographic.com/blog/c/islands/

KEY INTERNATIONAL CORAL REEF CONSERVATION GROUPS

The Nature Conservancy
http://www.nature.org/

Conservation International
http://www.conservation.org/Pages/default.aspx

World Wildlife Fund
http://www.worldwildlife.org/

RESEARCH PUBLICATIONS

Introduction
Hughes, T.P., Baird, A.H., Bellwood, D.R., Card, M., Connolly, S.R., Folke, C. et al. (2003) Climate change, human impacts, and the resilience of coral reefs. *Science* **301**: 929-933.

Knowlton, N. and J. B. C. Jackson (2008). Shifting baselines, local impacts, and global change on coral reefs. *PLoS Biology* **6**(2): 215-220.

Pandolfi, J.M., Bradbury, R.H., Sala, E., Hughes, T.P., Bjorndal, K.A., Cooke, R.G. et al. (2003) Global trajectories of the long-term decline of coral reef ecosystems. *Science* **301**: 955-958.

Chapter I, How to Build a Coral Reef
Falkowski, P.G., Dubinsky, Z., Muscatine, L., and Porter, J.W. (1984) Light and the bioenergetics of a symbiotic coral. *BioScience* **34**: 705-709.

Muscatine, L., and Porter, J.W. (1977) Reef corals: Mutualistic symbioses adapted to nutrient-poor environments. *BioScience* **27**: 454-460.

Odum, H.T., and Odum, E.P. (1955) Trophic structure and productivity of a windward coral reef community on Eniwetok Atoll. *Ecological Monographs* **25**: 291-320.

Chapter II, Global Stressors

Anthony, K.R.N., Kline, D.I., Diaz-Pulido, G., Dove, S., and Hoegh-Guldberg (2008) Ocean acidification causes bleaching and productivity loss in coral reef builders. *Proceedings of the National Academy of Sciences USA* **105**: 17442-17446.

Baker, A.C., Starger, C.J., McClanahan, T.R., and Glynn, P.W. (2004) Corals' adaptive response to climate change. *Nature* **430**: 741.

Brown, B.E. (1997) Coral bleaching: Causes and consequences. *Coral Reefs* **16**: S129-S138.

Buddemeier, R.W., and Fautin, D.G. (1993) Coral bleaching as an adaptive mechanism: A testable hypothesis. *BioScience* 43: 320-326.

Hoegh-Guldberg, O., Mumby, P.J., Hooten, A.J., Steneck, R.S., Greenfield, P., Gomez, E. et al. (2007) Coral reefs under rapid climate change and ocean acidification. *Science* **318**: 1737-1742.

Jones, R.J. (2008) Coral bleaching, bleaching-induced mortality, and the adaptive significance of the bleaching response. *Marine Biology* **154**: 65-80.

Rowan, R., and Knowlton, N. (1995) Intraspecific diversity and ecological zonation in coral-algal symbiosis. *Proceedings of the National Academy of Sciences USA* **92**: 2850-2853.

Toller, W.W., Rowan, R., and Knowlton, N. (2001) Repopulation of zooxanthellae in the Caribbean corals *Montastraea annularis* and *M. faveolata* following experimental and disease-associated bleaching. *Biological Bulletin* **201**: 360-373.

Chapter III, The Holobiont's Microbial Side

Johnston, I.S., and Rohwer, F. (2007) Microbial landscapes on the outer tissue surfaces of the reef-building coral *Porites compressa*. *Coral Reefs* **26**: 375-383.

Kellogg, C.A. (2004) Tropical Archaea: Diversity associated with the surface microlayer of corals. *Marine Ecology Progress Series* **273**: 81-88.

Knowlton, N., and Rohwer, F. (2003) Multispecies microbial mutualisms on coral reefs: The host as a habitat. *The American Naturalist* **162**: S51-S62.

Rohwer, F., Seguritan, V., Azam, F., and Knowlton, N. (2002) Diversity and distribution of coral-associated bacteria. *Marine Ecology Progress Series* **243**: 1-10.

Wegley, L., Yu, Y., Breitbart, M., Casas, V., Kline, D.I., and Rohwer, F. (2004) Coral-associated Archaea. *Marine Ecology Progress Series* **273**: 89-96.

Chapter IV, Coral Diseases
Ben-Haim, Y., Zicherman-Keren, M., and Rosenberg, E. (2003) Temperature-regulated bleaching and lysis of the coral *Pocillopora damicornis* by the novel pathogen *Vibrio coralliilyticus*. *Applied and Environmental Microbiology* **69**: 4236-4242.

Bourne, D.G., Garren, M., Work, T.M., Rosenberg, E., Smith, G.W., and Harvell, C.D. (2009) Microbial disease and the coral holobiont. *Trends in Microbiology* **17**: 554-562.

Casas, V., Kline, D.I., Wegley, L., Yu, Y., Breitbart, M., and Rohwer, F. (2004) Widespread association of a Rickettsiales-like bacterium with reef-building corals. *Environmental Microbiology* **6**: 1137-1148.

Cooney, R.P., Pantos, O., Le Tissier, M.D.A., Barer, M.R., O'Donnell, A.G., and Bythell, J.C. (2002) Characterization of the bacterial consortium associated with black band disease in coral using molecular microbiological techniques. *Environmental Microbiology* **4**: 401-413.

Efrony, R., Loya, Y., Bacharach, E., and Rosenberg, E. (2007) Phage therapy of coral disease. *Coral Reefs* **26**: 7-13.

Green, E.P., and Bruckner, A.W. (2000) The significance of coral disease epizootiology for coral reef conservation. *Biological Conservation* **96**: 347-361.

Harvell, C.D., Jordán-Dahlgren, E., Merkel, S., Rosenberg, E., Raymundo, L., Smith, G.W. et al. (2007) Coral disease, environmental drivers, and the balance between coral and microbial associates. *Oceanography* **20**: 172-195.

Krediet, C.J., Ritchie, K.B., Cohen, M., Lipp, E.K., Sutherland, K.P., and Teplitski, M. (2009) Utilization of mucus from the coral *Acropora palmata* by the pathogen *Serratia marcescens* and by environmental and coral commensal bacteria. *Applied and Environmental Microbiology* **75**: 3851-3858.

Kushmaro, A., Loya, Y., Fine, M., and Rosenberg, E. (1996) Bacterial infection and coral bleaching. *Nature* **380**: 396.

Lesser, M.P., Bythell, J.C., Gates, R.D., Johnstone, R.W., and Hoegh-Guldberg, O. (2007) Are infectious diseases really killing corals? Alternative interpretations of the experimental and ecological data. *Journal of Experimental Marine Biology and Ecology* **346**: 36-44.

Richardson, L.L. (1998) Coral diseases: What is really known? *Trends in Ecology and Evolution* **13**: 438-443.

Richardson, L.L., Smith, G.W., Ritchie, K.B., and Carlton, R.G. (2001) Integrating microbiological, microsensor, molecular, and physiologic techniques in the study of coral disease pathogenesis. *Hydrobiologia* **460**: 71-89.

Rosenberg, E., and Ben-Haim, Y. (2002) Microbial diseases of corals and global warming. *Environmental Microbiology* **4**: 318-326.

Teplitski, M., and Ritchie, K.B. (2009) How feasible is the biological control of coral diseases? *Trends in Ecology and Evolution* **24**: 378-385.

Thurber, R.V., Willner-Hall, D., Rodriguez-Mueller, B., Desnues, C., Edwards, R.A., Angly, F. et al. (2009) Metagenomic analysis of stressed coral holobionts. *Environmental Microbiology* **11**: 2148-2163.

Thurber, R.L.V., Barott, K.L., Hall, D., Liu, H., Rodriguez-Mueller, B., Desnues, C. et al. (2008) Metagenomic analysis indicates that stressors induce production of herpes-like viruses in the coral *Porites compressa*. *Proceedings of the National Academy of Sciences USA* **105**: 18413-18418.

Wilson, W.H., Dale, A.L., Davy, J.E., and Davy, S.K. (2005) An enemy within? Observations of virus-like particles in reef corals. *Coral Reefs* **24**: 145-148.

Chapter V, Overfishing and the Rise of the Algae

Beisner, B.E., Haydon, D.T., and Cuddington, K. (2003) Alternative stable states in ecology. *Frontiers in Ecology and the Environment* **1**: 376-382.

Gardner, T.A., Côté, I.M., Gill, J.A., Grant, A., and Watkinson, A.R. (2005) Hurricanes and Caribbean coral reefs: Impacts, recovery patterns, and role in long-term decline. *Ecology* **86**: 174–184.

Hughes, T.P. (1994) Catastrophes, phase shifts, and large-scale degradation of a Caribbean coral reef. *Science* **265**: 1547-1551.

Jackson, J.B.C. (1997) Reefs since Columbus. *Coral Reefs* **16**: S23-S32.

Jackson, J.B.C., Kirby, M.X., Berger, W.H., Bjorndal, K.A., Botsford, L.W., Bourque, B.J. et al. (2001) Historical overfishing and the recent collapse of coastal ecosystems. *Science* **293**: 629-638.

Knowlton, N. (1992) Thresholds and multiple stable states in coral reef community dynamics. *American Zoologist* **32**: 674-682.

Knowlton, N. (2001) The future of coral reefs. *Proceedings of the National Academy of Sciences USA* **98**: 5419-5425.

Nyström, M., Folke, C., and Moberg, F. (2000) Coral reef disturbance and resilience in a human-dominated environment. *Trends in Ecology and Evolution* **15**: 413-417.

Scheffer, M., Carpenter, S., Foley, J.A., Folke, C., and Walker, B. (2001) Catastrophic shifts in ecosystems. *Nature* **413**: 591-596.

Chapter VI, The DDAMnation of Coral Reefs
Azam, F., Fenchel, T., Field, J.G., Gray, J.S., Meyer-Reil, L.A., and Thingstad, F. (1983) The ecological role of water-column microbes in the sea. *Marine Ecology Progress Series* **10**: 257-263.

Azam, F. (1998) Microbial control of oceanic carbon flux: The plot thickens. *Science* **280**: 694-696.

Barott, K., Smith, J., Dinsdale, E., Hatay, M., Sandin, S., and Rohwer, F. (2009) Hyperspectral and physiological analyses of coral-algal interactions. *PLoS ONE* **4**: e8043.

Kline, D.I., Kuntz, N.M., Breitbart, M., Knowlton, N., and Rohwer, F. (2006) Role of elevated organic carbon levels and microbial activity in coral mortality. *Marine Ecology Progress Series* **314**: 119-125.

Koop, K., Booth, D., Broadbent, A., Brodie, J., Bucher, D., Capone, D. et al. (2001) ENCORE: The effect of nutrient enrichment on coral reefs. Synthesis of results and conclusions. *Marine Pollution Bulletin* **42**: 91-120.

Kuntz, N.M., Kline, D.I., Sandin, S.A., and Rohwer, F. (2005) Pathologies and mortality rates caused by organic carbon and nutrient stressors in three Caribbean coral species. *Marine Ecology Progress Series* **294**: 173-180.

McClanahan, T.R., Sala, E., Stickels, P.A., Cokos, B.A., Baker, A.C., Starger, C.J., and Jones IV, S.H. (2003) Interaction between nutrients and herbivory in controlling algal communities and coral condition on Glover's Reef, Belize. *Marine Ecology Progress Series* **261**: 135-147.

McCook, L.J., Jompa, J., and Diaz-Pulido, G. (2001) Competition between corals and algae on coral reefs: A review of evidence and mechanisms. *Coral Reefs* **19**: 400-417.

Miller, M.W., Hay, M.E., Miller, S.L., Malone, D., Sotka, E.E., and Szmant, A.M. (1999) Effects of nutrients versus herbivores on reef algae: A new method for manipulating nutrients on coral reefs. *Limnology and Oceanography* **44**: 1847-1861.

Smith, J.E., Shaw, M., Edwards, R.A., Obura, D., Pantos, O., Sala, E. et al. (2006) Indirect effects of algae on coral: Algae-mediated, microbe-induced coral mortality. *Ecology Letters* **9**: 834-845.

Chapter VII, More Nutrients Equals Even More Algae
Diaz, R.J., and Rosenberg, R. (2008) Spreading dead zones and consequences for marine ecosystems. *Science* **321**: 926-929.

Hunter, C.L., and Evans, C.W. (1995) Coral reefs in Kaneohe Bay, Hawaii: Two centuries of western influence and two decades of data. *Bulletin of Marine Science* **57**: 501-515.

Koop, K., Booth, D., Broadbent, A., Brodie, J., Bucher, D., Capone, D. et al. (2001) ENCORE: The effect of nutrient enrichment on coral reefs. Synthesis of results and conclusions. *Marine Pollution Bulletin* **42**: 91-120.

Loya, Y. (2007) How to influence environmental decision makers? The case of Eilat (Red Sea) coral reefs. *Journal of Experimental Marine Biology and Ecology* **344**: 35-53.

McClanahan, T.R., Sala, E., Stickels, P.A., Cokos, B.A., Baker, A.C., Starger, C.J., and Jones IV, S.H. (2003) Interaction between nutrients and herbivory in controlling algal communities and coral condition on Glover's Reef, Belize. *Marine Ecology Progress Series* **261**: 135-147.

McCook, L.J., Jompa, J., and Diaz-Pulido, G. (2001) Competition between corals and algae on coral reefs: A review of evidence and mechanisms. *Coral Reefs* **19**: 400-417.

Miller, M.W., Hay, M.E., Miller, S.L., Malone, D., Sotka, E.E., and Szmant, A.M. (1999) Effects of nutrients versus herbivores on reef algae: A new method for manipulating nutrients on coral reefs. *Limnology and Oceanography* **44**: 1847-1861.

Szmant, A.M. (2002) Nutrient enrichment on coral reefs: Is it a major cause of coral reef decline? *Estuaries* **25**: 743-766.

Vitousek, P.M., Aber, J.D., Howarth, R.W., Likens, G.E., Matson, P.A., Schindler, D.W. et al. (1997) Human alteration of the global nitrogen cycle: Sources and consequences. *Ecological Applications* **7**: 737-750.

Chapter VIII, The Microbialization of Christmas Atoll
Dinsdale, E.A., Pantos, O., Smriga, S., Edwards, R.A., Angly, F., Wegley, L. et al. (2008) Microbial ecology of four coral atolls in the Northern Line Islands. *PLoS One* **3**: e1584.

Friedlander, A.M., and DeMartini, E.E. (2002) Contrasts in density, size, and biomass of reef fishes between the northwestern and the main Hawaiian islands: The effects of fishing down apex predators. *Marine Ecology Progress Series* **230**: 253-264.

Maragos, J., Friedlander, A.M., Godwin, S., Musburger, C., Tsuda, R., Flint, E. et al. (2008) US Coral Reefs in the Line and Phoenix Islands, Central Pacific Ocean: Status, threats and significance. *In Coral Reefs of the USA*. Riegl, B.M., and Dodge, R.E. (eds): Springer.

Sandin, S.A., Smith, J.E., DeMartini, E.E., Dinsdale, E.A., Donner, S.D., Friedlander, A.M. et al. (2008) Baselines and degradation of coral reefs in the Northern Line Islands. *PLoS One* **3**: e1548.

Stevenson, C., Katz, L.S., Micheli, F., Block, B., Heiman, K.W., Perle, C. et al. (2007) High apex predator biomass on remote Pacific islands. *Coral Reefs* **26**: 47-51.

Chapter IX, Giving Coral Reefs a Chance

Branch, T.A. (2009) How do individual transferable quotas affect marine ecosystems? *Fish and Fisheries* **10**: 39-57.

Bromley, D.W. (2005) Purging the frontier from our mind: Crafting a new fisheries policy. *Reviews in Fish Biology and Fisheries* **15**: 217-229.

Heal, G., and Schlenker, W. (2008) Sustainable fisheries. *Nature* **455**: 1044-1045.

Hughes, T.P., Baird, A.H., Bellwood, D.R., Card, M., Connolly, S.R., Folke, C. et al. (2003) Climate change, human impacts, and the resilience of coral reefs. *Science* **301**: 929-933.

Hughes, T.P., Bellwood, D.R., Folke, C., Steneck, R.S., and Wilson, J. (2005) New paradigms for supporting the resilience of marine ecosystems. *Trends in Ecology and Evolution* **20**: 380-386.

Jameson, S.C. (2008) Reefs in trouble – The real root cause. *Marine Pollution Bulletin* **56**: 1513-1514.

Normile, D. (2009) Bringing coral reefs back from the living dead. *Science* **325**: 559-561.

Pandolfi, J.M., Jackson, J.B.C., Baron, N., Bradbury, R.H., Guzman, H.M., Hughes, T.P. et al. (2005) Are U.S. coral reefs on the slippery slope to slime? *Science* **307**: 1725-1726.

Paul, J.H., Rose, J.B., Brown, J., Shinn, E.A., Miller, S., and Farrah, S.R. (1995) Viral tracer studies indicate contamination of marine waters by sewage disposal practices in Key Largo, Florida. *Applied and Environmental Microbiology* **61**: 2230-2234.

Pauly, D., Christensen, V., Guénette, S., Pitcher, T.J., Sumaila, U.R., Walters, C.J. et al. (2002) Towards sustainability in world fisheries. *Nature* **418**: 689-695.

Sagarin, R.D., and Crowder, L.B. (2008) Breaking through the crisis in marine conservation and management: Insights from the philosophies of Ed Ricketts. *Conservation Biology* **23**: 24-30.

ACKNOWLEDGMENTS

A phenomenal amount of effort by many people went into this book and the experiences describe therein. First, I must thank the amazing people in my lab at **San Diego State University**. Without you, there is no way that any of this would have happened. The ones most closely associated with the coral project have been *Linda Wegley Kelly, Mya Breitbart, Matt Haynes, Rebecca Vega Thurber, Katie Barott, Florent Angly, Tracey McDole, Veronica Casas, Olga Pantos, Neilan Kuntz, Morrigan Shaw, Mike Furlan, Beltran Rodriguez-Mueller, Dana Willner, Jon Miyake, Christelle Duesnes, Jennifer Rodriguez-Mueller, Selina Liu, and Bahador Nosrat*. Also essential was Rebecca's Coffee House as a source of caffeine for much of the art work and writing.

I also want to thank my colleagues at San Diego State University who did much of the work. In particular, *Rob Edwards, Liz Dinsdale, Anca Segall, Stan Maloy, Kathie McGuire*, and *Constantine Tsoukas,* as well as the Math Group: *Peter Salamon, Ben Felts*, and *James Nulton*, and, of course, *Mark Hatay* who built all of our fancy tools and then helped put them to good use. I deeply appreciate the support of the greater San Diego scientific community, especially the long standing collaboration between SDSU and the **Scripps Institution of Oceanography**. Among those at SIO, I want to expressly acknowledge *Stuart Sandin, Jennifer Smith, Farooq Azam, Nancy Knowlton, Davey Kline, Kristen Marhaver, Jeremy Jackson, Wes Toller, Enric Sala*, and the Coral Club participants. A thank you to Fernando Nosratpour of the Birch Aquarium for supplying us with corals and helping us keep them alive. Other colleagues around the world who have been especially helpful include *Rusty Brainard, Rick Bushman, John Bythell, Craig Carlson, Doug Conrad, Ed Delong, Mohamed Fairoz Mohamed Farook, Bruce Fouke, Ian Johnston, Chris Kellogg, Ariel Kushmaro, Fred Lipshultz, Yossi Loya, Nancy Moran, Howard Ochman, John Paul, Paul Rainey, Kim Ritchie, Eugene Rosenberg, John Roth, Garriet Smith, Robbie Smith, Ernesto Weil, Ruan Yijun,* and *Dave Zawada. Jeffrey Gordon* and *David Relman,* thank you for exploring the human holobiont.

All onboard the expeditions know how much we owe—including our very lives—to *Captain Vincent Backen* of the White Holly and crew members *Uncle Bert Dorio, Matthew Guanci, Joanne Keune, Charles Kithcart, David Murphy,* and *Cody-the-Cook Reynolds.*

Speaking of those onboard, they were:

The Fish—*Ed DeMartini, Alan Friedlander, Enric Sala,* and *Stuart Sandin*

The Benthics—*Liz Dinsdale, Nancy Knowlton, Jim Maragos, David Obura, Gustav Paulay,* and *Jennifer Smith*

The Microbes—*Rob Edwards, Olga Pantos, Forest Rohwer,* and *Steven Smriga*

The Dive Safety Officers—*Christian McDonald* and *Mike Lang*

Thanks also to the participants on the Southern Line Islands cruise. We hope to get the data written up soon.

This venture was made possible by the financial support of the *National Science Foundation,* the *Gordon and Betty Moore Foundation,* the *Moore Family Foundation,* the *Agouron Institute, Conservation International,* the *Canadian Institute for Advanced Research,* and the *National*

Oceanic and Atmospheric Administration. We are beholden to numerous people at those funding agencies for their patient and diligent assistance. We also gratefully acknowledge the additional support provided by private donors. Thank you to Gina Spidel and Leslie Rodelander at the BioPod for helping us spend the money.

Thank you to everyone who reviewed the book in its various iterations: *Liz Dinsdale, Linda Wegley Kelly, Neilan Kuntz, Pat and Gary Rohwer, Anca Segall, Stuart Sandin*, and *Dave Zawada*.

Finally, appreciation beyond words to my family: at the top of the list, *Anca Segall*, who has contemplated (but never actually carried out) murdering me for disappearing for long periods of time; *Pat & Gary Rohwer* who, being my parents, can never escape; my grandmother, *Pat Smith*, who loves coral reefs as much as I do, and *Willow the Troublemaker*, who I hope will.

INDEX

About the
Authors and Artist